T0233040

Lecture Notes in Computer Science   8449

Commenced Publication in 1973
Founding and Former Series Editors:
Gerhard Goos, Juris Hartmanis, and Jan van Leeuwen

More information about this series at http://www.springer.com/series/7151

James F. Peters · Andrzej Skowron
Tianrui Li · Yan Yang
JingTao Yao · Hung Son Nguyen (Eds.)

Transactions on Rough Sets XVIII

 Springer

Editors-in-Chief

James F. Peters
ECE Department Computational
    Intelligence Laboratory
University of Manitoba
Winnipeg, MB
Canada

Andrzej Skowron
University of Warsaw
Warsaw
Poland

Guest Editors

Tianrui Li
Yan Yang
Southwest Jiaotong University
Sichuan
China

Hung Son Nguyen
University of Warsaw
Warsaw
Poland

JingTao Yao
University of Regina
Regina, SK
Canada

ISSN 0302-9743   ISSN 1611-3349   (electronic)
ISBN 978-3-662-44679-9   ISBN 978-3-662-44680-5   (eBook)
DOI 10.1007/978-3-662-44680-5

Library of Congress Control Number: 2014947523

Springer Heidelberg New York Dordrecht London

Printed on acid-free paper

Springer is part of Springer Science+Business Media (www.springer.com)

Preface

Volume XVIII of the *Transactions on Rough Sets* (TRS) is a continuation of a number of research streams that have grown out of the seminal work of Zdzisław Pawlak[1] during the first decade of the twenty-first century. This special issue is dedicated to the 2012 Joint Rough Set Symposium (JRS 2012) held in Chengdu, China, during August 17–20, 2012. JRS 2012 is a joint conference comprising the 8th International Conference on Rough Sets and Current Trends in Computing (RSCTC 2012) and the 7th International Conference on Rough Sets and Knowledge Technology (RSKT 2012). After peer review, seven extended papers were accepted for further revision. The papers reflect the current development of rough sets and three-way decisions.

The paper coauthored by You-Hong Xu, Wei-Zhi Wu, and Guoyin Wang investigates topological structures of rough intuitionistic fuzzy sets. The paper coauthored by Jiabin Liu, Fan Min, Hong Zhao, and William Zhu defines feature selection with positive region constraint for test-cost-sensitive data problems and develops a heuristic algorithm to deal with it. The paper coauthored by Singh Kavita, Zaveri Mukesh, and Raghuwanshi Mukesh develops a rough neurocomputing recognition system for illumination invariant face recognition. The paper coauthored by Hengrong Ju, Xibei Yang, Huili Dou, and Jingjing Song presents variable precision multigranulation rough sets and attribute reduction. The paper coauthored by Xiuyi Jia and Lin Shang examines a three-way decisions solution and a two-way decisions solution for filtering spam emails. The paper coauthored by Hong Yu, Ying Wang, and Peng Jiao provides a three-way decisions approach for overlapping clustering based on the decision-theoretic rough set model and develops a density-based overlapping clustering algorithm sing three-way decisions. The paper coauthored by Dun Liu, Tianrui Li, and Decui Liang presents a model of stochastic decision-theoretic rough sets and its properties together with a case study.

The editors of the special issue wish to express their gratitude to Professors James F. Peters and Andrzej Skowron, Editors-in-Chief, for agreeing to publish this special issue and for their guidance during the preparation of this special issue.

The editors would like to express gratitude to the authors of all submitted papers. Special thanks are due to the following reviewers: Chuan Luo, Shaoyong Li, Lin Shang, Shuang An, Pawan Lingras, Shoji Hirano, Xibei Yang, Xun Gong, Ling Wei, Jifang Pang, Fan Min, and Hong Yu.

---

[1] See, *e.g.*, Pawlak, Z., A Treatise on Rough Sets, *Transactions on Rough Sets* IV, (2006), 1–17. See, also, Pawlak, Z., Skowron, A.: Rudiments of rough sets, *Information Sciences* 177 (2007) 3–27; Pawlak, Z., Skowron, A.: Rough sets: Some extensions, *Information Sciences* 177 (2007) 28–40; Pawlak, Z., Skowron, A.: Rough sets and Boolean reasoning, *Information Sciences* 177 (2007) 41–73.

The editors and authors of this volume extend their gratitude to Alfred Hofmann, Anna Kramer, Ursula Barth, Christine Reiss, and the LNCS staff at Springer for their support in making this volume of the TRS possible.

The Editors-in-Chief were supported by the Polish National Science Centre (NCN) grants DEC-2011/01/B/ ST6/03 867, DEC-2011/01/D /ST6/ 06981, DEC-2012/05/B/ ST6/03215, DEC-2013/09/B/ST6/01568 as well as by the Polish National Centre for Research and Development (NCBiR) under the grant SYNAT No. SP/I/1/ 77065/10 in the framework of the strategic scientific research and experimental development program "Interdisciplinary System for Interactive Scientific and Scientific-Technical Information," the grants O ROB/0010/03/001, PBS2/B9/20/2013 as well as the Natural Sciences and Engineering Research Council of Canada (NSERC) discovery grant 185986.

April 2014

Tianrui Li
Yan Yang
JingTao Yao
Hung Son Nguyen
James F. Peters
Andrzej Skowron

LNCS Transactions on Rough Sets

The *Transactions on Rough Sets* series has as its principal aim the fostering of professional exchanges between scientists and practitioners who are interested in the foundations and applications of rough sets. Topics include foundations and applications of rough sets as well as foundations and applications of hybrid methods combining rough sets with other approaches important for the development of intelligent systems. The journal includes high-quality research articles accepted for publication on the basis of thorough peer reviews. Dissertations and monographs up to 250 pages that include new research results can also be considered as regular papers. Extended and revised versions of selected papers from conferences can also be included in regular or special issues of the journal.

Editors-in-Chief

James F. Peters
Andrzej Skowron

Managing Editor

Sheela Ramanna

Technical Editor

Marcin Szczuka

Editorial Board

Mohua Banerjee
Jan Bazan
Gianpiero Cattaneo
Mihir K. Chakraborty
Davide Ciucci
Chris Cornelis
Ivo Düntsch
Anna Gomolińska
Salvatore Greco
Jerzy W. Grzymała-Busse

Masahiro Inuiguchi
Jouni Järvinen
Richard Jensen
Bożena Kostek
Churn-Jung Liau
Pawan Lingras
Victor Marek
Mikhail Moshkov
Hung Son Nguyen
Ewa Orłowska

Contents

On the Intuitionistic Fuzzy Topological Structures of Rough Intuitionistic Fuzzy Sets

You-Hong Xu[1,2], Wei-Zhi Wu[1,2(✉)], and Guoyin Wang[2]

[1] School of Mathematics, Physics and Information Science,
Zhejiang Ocean University, Zhoushan 316022,
Zhejiang, People's Republic of China
{xyh,wuwz}@zjou.edu.cn
[2] Chongqing Key Laboratory of Computational Intelligence,
Chongqing University of Posts and Telecommunications,
Chongqing 400065, People's Republic of China
wanggy@ieee.org

Abstract. A rough intuitionistic fuzzy set is the result of approximation of an intuitionistic fuzzy set with respect to a crisp approximation space. In this paper, we investigate topological structures of rough intuitionistic fuzzy sets. We first show that a reflexive crisp rough approximation space can induce an intuitionistic fuzzy Alexandrov space. It is proved that the lower and upper rough intuitionistic fuzzy approximation operators are, respectively, an intuitionistic fuzzy interior operator and an intuitionistic fuzzy closure operator if and only if the binary relation in the crisp approximation space is reflexive and transitive. We then verify that a similarity crisp approximation space can produce an intuitionistic fuzzy clopen topological space. We further examine sufficient and necessary conditions that an intuitionistic fuzzy interior (closure, respectively) operator derived from an intuitionistic fuzzy topological space can associate with a reflexive and transitive crisp relation such that the induced lower (upper, respectively) rough intuitionistic fuzzy approximation operator is exactly the intuitionistic fuzzy interior (closure, respectively) operator.

Keywords: Approximation operators · Binary relations · Intuitionistic fuzzy sets · Intuitionistic fuzzy topologies · Rough intuitionistic fuzzy sets · Rough sets

1 Introduction

One of the main directions in the research of rough set theory [31] is naturally the generalization of concepts of Pawlak's rough set approximation operators. Many authors generalized Pawlak's rough set approximations by using non-equivalence relations. It is well-known that data with fuzzy set values are commonly seen in real-world applications and fuzzy-set concepts are often used to represent quantitative data expressed in linguistic terms and membership

© Springer-Verlag Berlin Heidelberg 2014
J.F. Peters et al. (Eds.): Transactions on Rough Sets XVIII, LNCS 8449, pp. 1–22, 2014.
DOI: 10.1007/978-3-662-44680-5_1

functions in intelligent systems. Based on this observation, many authors have generalized the Pawlak's rough set model to the fuzzy environment. The results of these studies lead to the introduction of notions of rough fuzzy sets (fuzzy sets approximated by a crisp approximation space) [14, 46, 49] and fuzzy rough sets (fuzzy or crisp sets approximated by a fuzzy approximation space) [14, 28–30, 35, 43–45, 49]. The rough fuzzy set model may be used to handle knowledge acquisition in information systems with fuzzy decisions while the fuzzy rough set model may be employed to unravel knowledge hidden in fuzzy decision systems.

As a more general case of fuzzy sets, the concept of intuitionistic fuzzy (IF for short) sets, which was originated by Atanassov [3], has played a useful role in the research of uncertainty theories. Unlike a fuzzy set, which gives a degree of which element belongs to a set, an IF set gives both a membership degree and a nonmembership degree. Obviously, an IF set is more objective than a fuzzy set to describe the vagueness of data or information. Therefore, the combination of IF set theory and rough set theory is a new hybrid model to describe the uncertain information and is an interesting research issue over the years (see e.g. [5, 11, 12, 16, 36, 37, 47, 54, 56]). The resulting rough sets in the IF environment are indeed natural generalizations of Pawlak's original concept of rough sets and will be useful for the analysis of vague data sets and interval-valued data sets.

Topology is a branch of mathematics, whose concepts exist not only in almost all branches of mathematics, but also in many real life applications. Topologies are widely used in the research field of data mining and knowledge discovery (see e.g. [1, 7, 17, 20, 21, 24, 25]). The concept of topological structures and their generalizations are powerful notions and are important bases in data and system analysis. An interesting and important research for rough approximation operators is to compare them with the topological properties and structures. Many authors studied topological structures of rough sets in the literature. For example, Chuchro [8,9], Kondo [18], Lashin *et al.* [23], li *et al.* [26], Pei *et al.* [32], Qin *et al.* [34], Wiweger [40], Wu *et al.* [41], Yang and Xu [51], and Zhu [57] studied the topological structures for crisp rough sets. Boixader *et al.* [4], Hao and Li [15], Kortelainen [19], Qin and Pei [33], Thiele [38], Tiwari and Srivastava [39], Wu *et al.* [42, 43, 45, 48], respectively, discussed topological structures of rough sets in the fuzzy environment. One of the main results is that a reflexive and transitive approximation space can yield a topology on the same universe such that the lower and upper approximation operators of the approximation space are, respectively, the interior and closure operators of the topology, and conversely, under some conditions, a topology can be associated with a reflexive and transitive approximation space which produces the same topology. There exists a one-to-one correspondence between the set of all reflexive, transitive relations and the set of Alexandrov topologies on an arbitrary universe [15, 22, 33, 39, 43, 48, 53].

Although the concept of IF topological space was defined in [10], compared with the topological structures of rough set in the fuzzy environment, lesser effort has been made on the study of topological structures of rough sets in the IF environment. Wu and Zhou [50], and Zhou and Wu [55] recently discussed the relationship between IF rough sets (the results of approximations of intuitionistic fuzzy sets with respect to an IF approximation space) and IF topological spaces and showed that for an IF reflexive and transitive approximation space there exists an IF topological space such that the lower and upper IF rough approximation operators induced from the IF approximation space are, respectively, the interior and closure operators in the IF topological space. They also examined the sufficient and necessary conditions that an IF interior (closure, respectively) operator derived from an IF topological space can associate with an IF reflexive and transitive relation such that the induced lower (upper, respectively) IF rough approximation operator is exactly the IF interior (closure, respectively) operator.

We know that a rough IF set is the result of approximation of an IF set with respect to a crisp approximation space. So, the rough IF set model is a typical rough set model in the IF environment. To improve and develop the applications of topology and rough sets on IF uncertain information, topological structures of rough IF sets need to be studied. Therefore, the main objective of the present paper is to investigate topological structures of rough IF sets. In the next section, we review basic notions of IF sets. In Sect. 3, we introduce concepts related to rough IF sets and present some properties of rough IF approximation operators. In Sect. 4, we give some notions and theoretical results of IF topological spaces. Section 5 examines IF topological structures of rough IF sets. In Sect. 6, we investigate under which conditions that an IF topology can be associated with a crisp approximation space which produces the same IF topology. We then conclude the paper with a summary in Sect. 7.

2   Basic Notions of Intuitionistic Fuzzy Sets

In this section, we introduce some basic notions of intuitionistic fuzzy sets. We first review a special lattice on $[0,1] \times [0,1]$ (where $[0,1]$ is the unit interval) originated by Cornelis *et al.* [13].

Throughout this paper, U will be a nonempty set called the universe of discourse. The class of all subsets (respectively, fuzzy subsets) of U will be denoted by $\mathcal{P}(U)$ (respectively, by $\mathcal{F}(U)$). For $y \in U$, 1_y will denote the fuzzy singleton with value 1 at y and 0 elsewhere; 1_M will denote the characteristic function of a set $M \in \mathcal{P}(U)$. For $\alpha \in [0,1]$, $\widehat{\alpha}$ will denote the constant fuzzy set: $\widehat{\alpha}(x) = \alpha$, for all $x \in U$. For any $A \in \mathcal{F}(U)$, the complement of A will be denoted by $\sim A$, i.e. $(\sim A)(x) = 1 - A(x)$ for all $x \in U$. We will use the symbols $\vee$ and $\wedge$ to denote the supremum and the infimum, respectively.

Definition 1. (Lattice $(L^*, \leq_{L^*})$). *Denote*

$$L^* = \{(x_1, x_2) \in [0,1] \times [0,1] \mid x_1 + x_2 \leq 1\}.$$

A relation $\leq_{L^*}$ on L^* is defined as follows: $\forall (x_1, x_2), (y_1, y_2) \in L^*$,

$$(x_1, x_2) \leq_{L^*} (y_1, y_2) \iff x_1 \leq y_1 \text{ and } x_2 \geq y_2. \tag{1}$$

The relation $\leq_{L^*}$ is a partial ordering on L^* and the pair $(L^*, \leq_{L^*})$ is a complete lattice with the smallest element $0_{L^*} = (0,1)$ and the greatest element

$1_{L^*} = (1,0)$ [13]. *The meet operator $\wedge$ and join operator $\vee$ on $(L^*, \leq_{L^*})$ which are linked to the ordering $\leq_{L^*}$ are, respectively, defined as follows:* $\forall (x_1, x_2)$, $(y_1, y_2) \in L^*$,

$$(x_1, x_2) \wedge (y_1, y_2) = (\min(x_1, y_1), \max(x_2, y_2)), \tag{2}$$

$$(x_1, x_2) \vee (y_1, y_2) = (\max(x_1, y_1), \min(x_2, y_2)). \tag{3}$$

Meanwhile, an order relation $\geq_{L^*}$ on L^* is defined as follows: $\forall x = (x_1, x_2)$, $y = (y_1, y_2) \in L^*$,

$$(y_1, y_2) \geq_{L^*} (x_1, x_2) \iff (x_1, x_2) \leq_{L^*} (y_1, y_2), \tag{4}$$

and

$$x = y \iff x \leq_{L^*} y \text{ and } y \leq_{L^*} x.$$

Definition 2. [3] *Let a set U be fixed. An intuitionistic fuzzy set A in U is an object having the form*

$$A = \{\langle x, \mu_A(x), \gamma_A(x) \rangle \mid x \in U\},$$

where the functions $\mu_A : U \to [0,1]$ and $\gamma_A : U \to [0,1]$ define the degree of membership and the degree of non-membership of the element $x \in U$ to A, respectively. The functions μ_A and γ_A satisfy: $\mu_A(x) + \gamma_A(x) \leq 1$ for all $x \in U$. The family of all IF subsets in U is denoted by $\mathcal{IF}(U)$. The complement of an IF set A is denoted by $\sim A = \{\langle x, \gamma_A(x), \mu_A(x) \rangle \mid x \in U\}$.

Formally, an IF set A associates with two fuzzy sets $\mu_A : U \to [0,1]$ and $\gamma_A : U \to [0,1]$ and can be represented as $A = (\mu_A, \gamma_A)$. Obviously, any fuzzy set $A = \mu_A = \{\langle x, \mu_A(x) \rangle \mid x \in U\}$ may be identified with the IF set in the form $A = \{\langle x, \mu_A(x), 1 - \mu_A(x) \rangle \mid x \in U\}$.

We now introduce operations on $\mathcal{IF}(U)$ as follows [3]: for $A, B, A_i \in \mathcal{IF}(U)$, $i \in J$, J is an index set,

- $A \subseteq B$ if and only if (iff) $\mu_A(x) \leq \mu_B(x)$ and $\gamma_A(x) \geq \gamma_B(x)$, $\forall x \in U$,
- $A \supseteq B$ iff $B \subseteq A$,
- $A = B$ iff $A \subseteq B$ and $B \subseteq A$,
- $A \cap B = \{\langle x, \min(\mu_A(x), \mu_B(x)), \max(\gamma_A(x), \gamma_B(x)) \rangle \mid x \in U\}$,
- $A \cup B = \{\langle x, \max(\mu_A(x), \mu_B(x)), \min(\gamma_A(x), \gamma_B(x)) \rangle \mid x \in U\}$,
- $\bigcap_{i \in J} A_i = \{\langle x, \bigwedge_{i \in J} \mu_{A_i}(x), \bigvee_{i \in J} \gamma_{A_i}(x) \rangle \mid x \in U\}$,
- $\bigcup_{i \in J} A_i = \{\langle x, \bigvee_{i \in J} \mu_{A_i}(x), \bigwedge_{i \in J} \gamma_{A_i}(x) \rangle \mid x \in U\}$.

Now we define some special IF sets: a constant IF set $\widehat{(\alpha, \beta)} = \{\langle x, \alpha, \beta \rangle \mid x \in U\}$, where $(\alpha, \beta) \in L^*$; the IF universe set is $U = 1_U = \widehat{(1,0)} = \widehat{1_{L^*}} = \{\langle x, 1, 0 \rangle \mid x \in U\}$ and the IF empty set is $\emptyset = 0_U = \widehat{(0,1)} = \widehat{0_{L^*}} = \{\langle x, 0, 1 \rangle \mid x \in U\}$.

For any $y \in U$, IF sets 1_y and $1_{U-\{y\}}$ are, respectively, defined by: $\forall x \in U$,

$$\mu_{1_y}(x) = \begin{cases} 1, & \text{if } x = y, \\ 0, & \text{if } x \neq y. \end{cases} \qquad \gamma_{1_y}(x) = \begin{cases} 0, & \text{if } x = y, \\ 1, & \text{if } x \neq y. \end{cases}$$

$$\mu_{1_{U-\{y\}}}(x) = \begin{cases} 0, & \text{if } x = y, \\ 1, & \text{if } x \neq y. \end{cases} \qquad \gamma_{1_{U-\{y\}}}(x) = \begin{cases} 1, & \text{if } x = y, \\ 0, & \text{if } x \neq y. \end{cases}$$

Thus we have

$$1_y(x) = (\mu_{1_y}(x), \gamma_{1_y}(x)) = \begin{cases} (1,0) = 1_{L^*}, & \text{if } x = y, \\ (0,1) = 0_{L^*}, & \text{if } x \neq y. \end{cases}$$

And

$$1_{U-\{y\}}(x) = (\mu_{1_{U-\{y\}}}(x), \gamma_{1_{U-\{y\}}}(x)) = \begin{cases} (0,1) = 0_{L^*}, & \text{if } x = y, \\ (1,0) = 1_{L^*}, & \text{if } x \neq y. \end{cases}$$

By using L^*, we can observe that IF sets on U can be represented as follows: for $A, B, A_j \in \mathcal{IF}(U)(j \in J, J$ is an index set), $x, y \in U$, and $M \in \mathcal{P}(U)$,

- $A(x) = (\mu_A(x), \gamma_A(x)) \in L^*$,
- $U(x) = (1,0) = 1_{L^*}$,
- $\emptyset(x) = (0,1) = 0_{L^*}$,
- $x = y \implies 1_y(x) = 1_{L^*}$ and $1_{U-\{y\}}(x) = 0_{L^*}$,
- $x \neq y \implies 1_y(x) = 0_{L^*}$ and $1_{U-\{y\}}(x) = 1_{L^*}$,
- $x \in M \implies 1_M(x) = 1_{L^*}$,
- $x \notin M \implies 1_M(x) = 0_{L^*}$,
- $A \subseteq B \iff A(x) \leq_{L^*} B(x), \forall x \in U \iff B(x) \geq_{L^*} A(x), \forall x \in U$,
- $\left(\bigcap_{j \in J} A_j \right)(x) = \bigwedge_{j \in J} A_j(x) = \left(\bigwedge_{j \in J} \mu_{A_j}(x), \bigvee_{j \in J} \gamma_{A_j}(x) \right) \in L^*$,
- $\left(\bigcup_{j \in J} A_j \right)(x) = \bigvee_{j \in J} A_j(x) = \left(\bigvee_{j \in J} \mu_{A_j}(x), \bigwedge_{j \in J} \gamma_{A_j}(x) \right) \in L^*$.

3   Rough Intuitionistic Fuzzy Sets

In this section, we introduce the concept of rough IF sets and present some essential properties of rough IF approximation operators. We first review the notion of rough fuzzy sets with their properties which were introduced in [46, 48, 49].

3.1   Rough Fuzzy Approximation Operators

Definition 3. *Let U and W be two nonempty universes of discourse which may be infinite. A subset $R \in \mathcal{P}(U \times W)$ is referred to as a (crisp) binary relation from U to W. The relation R is referred to as serial if for each $x \in U$ there exists $y \in W$ such that $(x,y) \in R$; If $U = W$, R is referred to as a binary relation on U. R is referred to as reflexive if for all $x \in U$, $(x,x) \in R$; R is referred to as symmetric if for all $x, y \in U$, $(x,y) \in R$ implies $(y,x) \in R$; R is referred to as transitive if for all $x, y, z \in U$, $(x,y) \in R$ and $(y,z) \in R$ imply $(x,z) \in R$; R is referred to as Euclidean if for all $x, y, z \in U, (x,y) \in R$ and $(x,z) \in R$ imply $(y,z) \in R$; R is referred to as a similarity relation if R is reflexive and symmetric; R is referred to as a preorder if R is reflexive and transitive; and R is referred to as an equivalence relation if R is reflexive, symmetric and transitive.*

For an arbitrary crisp relation R from U to W, we can define a set-valued mapping $R_s : U \to \mathcal{P}(W)$ by:

$$R_s(x) = \{y \in W | (x, y) \in R\}, \quad x \in U. \tag{5}$$

$R_s(x)$ is referred to as the successor neighborhood of x with respect to (w.r.t.) R.

A rough fuzzy set is the approximation of a fuzzy set w.r.t. a crisp approximation space [14, 49].

Definition 4. *Let U and W be two non-empty universes of discourse and R a crisp binary relation from U to W, then the triple (U, W, R) is called a generalized (crisp) approximation space. For any fuzzy set $A \in \mathcal{F}(W)$, the lower and upper approximations of A, $\underline{RF}(A)$ and $\overline{RF}(A)$, w.r.t. the approximation space (U, W, R) are fuzzy sets of U whose membership functions, for each $x \in U$, are, respectively, defined by*

$$\underline{RF}(A)(x) = \bigwedge_{y \in R_s(x)} A(y), \quad \overline{RF}(A)(x) = \bigvee_{y \in R_s(x)} A(y). \tag{6}$$

The pair $(\underline{RF}(A), \overline{RF}(A))$ is called a generalized rough fuzzy set, and $\underline{RF}$ and $\overline{RF} : \mathcal{F}(W) \to \mathcal{F}(U)$ are referred to as the lower and upper rough fuzzy approximation operators, respectively.

By Definition 4, we can obtain properties of rough fuzzy approximation operators [46, 48, 49].

Theorem 1. *The lower and upper rough fuzzy approximation operators, $\underline{RF}$ and $\overline{RF}$, defined by Eq. (6) satisfy the following properties: For all $A, B \in \mathcal{F}(W)$, $A_j \in \mathcal{F}(W)(\forall j \in J)$, J is an index set, $y \in W$, and $\alpha \in [0, 1]$,*

(FL1) $\underline{RF}(A) =\sim \overline{RF}(\sim A)$,   (FU1) $\overline{RF}(A) =\sim \underline{RF}(\sim A)$,

(FL2) $\underline{RF}(A \cup \widehat{\alpha}) = \underline{RF}(A) \cup \widehat{\alpha}$,   (FU2) $\overline{RF}(A \cap \widehat{\alpha}) = \overline{RF}(A) \cap \widehat{\alpha}$;

(FL3) $\underline{RF}(\bigcap_{j \in J} A_j) = \bigcap_{j \in J} \underline{RF}(A_j)$,   (FU3) $\overline{RF}(\bigcup_{j \in J} A_j) = \bigcup_{j \in J} \overline{RF}(A_j)$,

(FL4) $A \subseteq B \Longrightarrow \underline{RF}(A) \subseteq \underline{RF}(B)$, (FU4) $A \subseteq B \Longrightarrow \overline{RF}(A) \subseteq \overline{RF}(B)$,

(FL5) $\underline{RF}(\bigcup_{j \in J} A_j) \supseteq \bigcup_{j \in J} \underline{RF}(A_j)$,   (FU5) $\overline{RF}(\bigcap_{j \in J} A_j) \subseteq \bigcap_{j \in J} \overline{RF}(A_j)$.

(FLC) $\underline{RF}(1_{W-\{y\}}) \in \mathcal{P}(U)$,   (FUC) $\overline{RF}(1_y) \in \mathcal{P}(U)$.

Properties (FL1) and (FU1) show that the rough fuzzy approximation operators $\underline{RF}$ and $\overline{RF}$ are dual with each other. Properties with the same number may be regarded as dual properties. Properties (FL3) and (FU3) state that the lower rough fuzzy approximation operator $\underline{RF}$ is multiplicative, and the upper rough fuzzy approximation operator $\overline{RF}$ is additive. One may also say that $\underline{RF}$ is distributive w.r.t. the intersection of fuzzy sets, and $\overline{RF}$ is distributive w.r.t. the union of fuzzy sets. Properties (FL5) and (FU5) imply that $\underline{RF}$ is not distributive w.r.t. set union, and $\overline{RF}$ is not distributive w.r.t. set intersection. However, properties (FL2) and (FU2) show that $\underline{RF}$ is distributive w.r.t. the union of a fuzzy set and a constant fuzzy set, and $\overline{RF}$ is distributive w.r.t. the intersection of a fuzzy set and a constant fuzzy set. Evidently, properties (FL2) and (FU2) imply the following properties:

$$(\text{FL2})' \quad \underline{RF}(W) = U, \qquad (\text{FU2})' \quad \overline{RF}(\emptyset) = \emptyset.$$

Property (FLC) shows that the lower approximation of the complement of a fuzzy singleton is a crisp set, and dually, property (FUC) implies that the upper approximation of a fuzzy singleton is a crisp set.

Analogous to Yao's study in [52], a serial rough fuzzy set model is obtained from a serial binary relation. The property of a serial relation can be characterized by the properties of its induced rough fuzzy approximation operators [46,48,49].

Theorem 2. *If R is an arbitrary crisp relation from U to W, and $\underline{RF}$ and $\overline{RF}$ are the rough fuzzy approximation operators defined by Eq.(6), then*

$$
\begin{aligned}
R \text{ is serial} &\Longleftrightarrow (\text{FL0}) \quad \underline{RF}(\emptyset) = \emptyset, \\
&\Longleftrightarrow (\text{FU0}) \quad \overline{RF}(W) = U, \\
&\Longleftrightarrow (\text{FL0})' \quad \underline{RF}(\widehat{\alpha}) = \widehat{\alpha}, \qquad \forall \alpha \in [0,1], \\
&\Longleftrightarrow (\text{FU0})' \quad \overline{RF}(\widehat{\alpha}) = \widehat{\alpha}, \qquad \forall \alpha \in [0,1], \\
&\Longleftrightarrow (\text{FLU0}) \quad \underline{RF}(A) \subseteq \overline{RF}(A), \forall A \in \mathcal{F}(W).
\end{aligned}
$$

In the case of connections between other special crisp relations and rough fuzzy approximation operators, we have the following [46,48,49]

Theorem 3. *Let R be an arbitrary crisp relation on U, and $\underline{RF}$ and $\overline{RF}$ the lower and upper rough fuzzy approximation operators defined by Eq.(6). Then*

$$
\begin{aligned}
R \text{ is reflexive} &\Longleftrightarrow (\text{FLR}) \quad \underline{RF}(A) \subseteq A, \quad \forall A \in \mathcal{F}(U), \\
&\Longleftrightarrow (\text{FUR}) \quad A \subseteq \overline{RF}(A), \quad \forall A \in \mathcal{F}(U). \\
R \text{ is symmetric} &\Longleftrightarrow (\text{FLS}) \quad \overline{RF}(\underline{RF}(A)) \subseteq A, \quad \forall A \in \mathcal{F}(U), \\
&\Longleftrightarrow (\text{FUS}) \quad A \subseteq \underline{RF}(\overline{RF}(A)), \quad \forall A \in \mathcal{F}(U), \\
&\Longleftrightarrow (\text{FLS})' \quad \underline{RF}(1_{U-\{x\}})(y) = \underline{RF}(1_{U-\{y\}})(x), \quad \forall(x,y) \in U \times U, \\
&\Longleftrightarrow (\text{FUS})' \quad \overline{RF}(1_x)(y) = \overline{RF}(1_y)(x), \quad \forall(x,y) \in U \times U. \\
R \text{ is transitive} &\Longleftrightarrow (\text{FLT}) \quad \underline{RF}(A) \subseteq \underline{RF}(\underline{RF}(A)), \quad \forall A \in \mathcal{F}(U), \\
&\Longleftrightarrow (\text{FUT}) \quad \overline{RF}(\overline{RF}(A)) \subseteq \overline{RF}(A), \quad \forall A \in \mathcal{F}(U). \\
R \text{ is Euclidean} &\Longleftrightarrow (\text{FLE}) \quad \overline{RF}(\underline{RF}(A)) \subseteq \underline{RF}(A), \quad \forall A \in \mathcal{F}(U), \\
&\Longleftrightarrow (\text{FUE}) \quad \overline{RF}(A) \subseteq \underline{RF}(\overline{RF}(A)), \quad \forall A \in \mathcal{F}(U).
\end{aligned}
$$

3.2 Rough Intuitionistic Fuzzy Approximation Operators

A rough IF set is the approximation of an IF set w.r.t. a crisp approximation space.

Definition 5. *Let (U, W, R) be a crisp approximation space. For an IF set $A = \{\langle x, \mu_A(x), \gamma_A(x) \rangle \mid x \in W\} \in \mathcal{IF}(W)$, the lower and upper approximations of A w.r.t. (U, W, R), denoted by $\underline{RI}(A)$ and $\overline{RI}(A)$, respectively, are defined as follows:*

$$\underline{RI}(A) = \{\langle x, \mu_{\underline{RI}(A)}(x), \gamma_{\underline{RI}(A)}(x) \rangle \mid x \in U\}, \tag{7}$$

$$\overline{RI}(A) = \{\langle x, \mu_{\overline{RI}(A)}(x), \gamma_{\overline{RI}(A)}(x) \rangle \mid x \in U\}, \tag{8}$$

where

$$\mu_{\underline{RI}(A)}(x) = \bigwedge_{y \in R_s(x)} \mu_A(y), \quad \gamma_{\underline{RI}(A)}(x) = \bigvee_{y \in R_s(x)} \gamma_A(y). \tag{9}$$

$$\mu_{\overline{RI}(A)}(x) = \bigvee_{y \in R_s(x)} \mu_A(y), \quad \gamma_{\overline{RI}(A)}(x) = \bigwedge_{y \in R_s(x)} \gamma_A(y). \tag{10}$$

It is easy to verify that $\underline{RI}(A)$ and $\overline{RI}(A)$ are two IF sets in U, thus IF mappings $\underline{RI}, \overline{RI} : \mathcal{IF}(W) \to \mathcal{IF}(U)$ are referred to as the lower and upper rough IF approximation operators, respectively, and the pair $(\underline{RI}(A), \overline{RI}(A))$ is called the rough IF set of A w.r.t. the approximation space (U, W, R).

According to Definitions 4 and 5, it can be seen that

$$\begin{aligned} \mu_{\underline{RI}(A)} &= \underline{RF}(\mu_A), \quad \gamma_{\underline{RI}(A)} = \overline{RF}(\gamma_A), \\ \mu_{\overline{RI}(A)} &= \overline{RF}(\mu_A), \quad \gamma_{\overline{RI}(A)} = \underline{RF}(\gamma_A). \end{aligned} \tag{11}$$

Then

$$\underline{RI}(A) = (\mu_{\underline{RI}(A)}, \gamma_{\underline{RI}(A)}) = (\underline{RF}(\mu_A), \overline{RF}(\gamma_A)). \tag{12}$$

Similarly,

$$\overline{RI}(A) = (\mu_{\overline{RI}(A)}, \gamma_{\overline{RI}(A)}) = (\overline{RF}(\mu_A), \underline{RF}(\gamma_A)). \tag{13}$$

Remark 1. When $A \in \mathcal{F}(W)$, in such a case, $\mu_A(x) + \gamma_A(x) = 1$ for all $x \in W$, then it is easy to observe that $(\underline{RI}(A), \overline{RI}(A))$ is a rough fuzzy set [48].

The following Theorem 4 presents some basic properties of rough IF approximation operators [47].

Theorem 4. *Let (U, W, R) be a crisp approximation space, then the lower and upper rough IF approximation operators defined in Definition 5 satisfy the following properties: $\forall A, B, A_j \in \mathcal{IF}(W), j \in J, J$ is an index set, $\forall (\alpha, \beta) \in L^*$, and $\forall y \in W$,*

(IL1) $\underline{RI}(A) =\sim \overline{RI}(\sim A)$,
(IU1) $\overline{RI}(A) =\sim \underline{RI}(\sim A)$;
(IL2) $\underline{RI}(A \cup \widehat{(\alpha, \beta)}) = \underline{RI}(A) \cup \widehat{(\alpha, \beta)}$,
(IU2) $\overline{RI}(A \cap \widehat{(\alpha, \beta)}) = \overline{RI}(A) \cap \widehat{(\alpha, \beta)}$;
(IL2)$'$ $\underline{RI}(W) = U$,
(IU2)$'$ $\overline{RI}(\emptyset) = \emptyset$;
(IL3) $\underline{RI}(\bigcap_{j \in J} A_j) = \bigcap_{j \in J} \underline{RI}(A_j)$,

(IU3) $\overline{RI}(\bigcup_{j \in J} A_j) = \bigcup_{j \in J} \overline{RI}(A_j);$

(IL4) $\underline{RI}(\bigcup_{j \in J} A_j) \supseteq \bigcup_{j \in J} \underline{RI}(A_j),$

(IU4) $\overline{RI}(\bigcap_{j \in J} A_j) \subseteq \bigcap_{j \in J} \overline{RI}(A_j);$

(IL5) $A \subseteq B \Longrightarrow \underline{RI}(A) \subseteq \underline{RI}(B),$

(IU5) $A \subseteq B \Longrightarrow \overline{RI}(A) \subseteq \overline{RI}(B).$

(ILC) $\underline{RI}(1_{W-\{y\}}) \in \mathcal{P}(U),$

(IUC) $\overline{RI}(1_y) \in \mathcal{P}(U).$

Properties (IL1) and (IU1) in Theorem 4 show that $\underline{RI}$ and $\overline{RI}$ are dual approximation operators. Similar to the case of rough fuzzy approximation operators, Theorem 5 below shows that the property of a serial relation can be characterized by the properties of its induced rough IF approximation operators.

Theorem 5. [47] *Let* (U, W, R) *be a crisp approximation space, and* $\underline{RI}$ *and* $\overline{RI}$ *the rough IF approximation operators defined in Definition 5. Then*

R *is serial*

$\Longleftrightarrow$ (IL0) $\underline{RI}(\emptyset) = \emptyset,$

$\Longleftrightarrow$ (IU0) $\overline{RI}(W) = U,$

$\Longleftrightarrow$ (IL0)$'$ $\underline{RI}(\widehat{(\alpha, \beta)}) = \widehat{(\alpha, \beta)}, \forall (\alpha, \beta) \in L^*,$

$\Longleftrightarrow$ (IU0)$'$ $\overline{RI}(\widehat{(\alpha, \beta)}) = \widehat{(\alpha, \beta)}, \forall (\alpha, \beta) \in L^*,$

$\Longleftrightarrow$ (ILU0) $\underline{RI}(A) \subseteq \overline{RI}(A), \forall A \in \mathcal{IF}(W).$

In the case of connections between special types of crisp relations and properties of rough IF approximation operators, we have the following

Theorem 6. [47] *Let* R *be an arbitrary crisp binary relation on* U, *and* $\underline{RI}, \overline{RI}$: $\mathcal{IF}(U) \to \mathcal{IF}(U)$ *the lower and upper rough IF approximation operators. Then*

(1) R *is reflexive*

$\Longleftrightarrow$ (ILR) $\underline{RI}(A) \subseteq A, \quad \forall A \in \mathcal{IF}(U),$

$\Longleftrightarrow$ (IUR) $A \subseteq \overline{RI}(A), \quad \forall A \in \mathcal{IF}(U).$

(2) R *is symmetric*

$\Longleftrightarrow$ (ILS) $\overline{RI}(\underline{RI}(A)) \subseteq A, \quad \forall A \in \mathcal{IF}(U),$

$\Longleftrightarrow$ (IUS) $A \subseteq \underline{RI}(\overline{RI}(A)), \quad \forall A \in \mathcal{IF}(U),$

$\Longleftrightarrow$ (ILS)$'$ $\mu_{\underline{RI}(1_{U-\{x\}})}(y) = \mu_{\underline{RI}(1_{U-\{y\}})}(x), \quad \forall (x, y) \in U \times U,$

$\Longleftrightarrow$ (IUS)$'$ $\mu_{\overline{RI}(1_x)}(y) = \mu_{\overline{RI}(1_y)}(x), \quad \forall (x, y) \in U \times U,$

$\Longleftrightarrow$ (ILS)$''$ $\gamma_{\underline{RI}(1_{U-\{x\}})}(y) = \gamma_{\underline{RI}(1_{U-\{y\}})}(x), \quad \forall (x, y) \in U \times U,$

$\Longleftrightarrow$ (IUS)$''$ $\gamma_{\overline{RI}(1_x)}(y) = \gamma_{\overline{RI}(1_y)}(x), \quad \forall (x, y) \in U \times U.$

(3) R *is transitive*

$\Longleftrightarrow$ (ILT) $\underline{RI}(A) \subseteq \underline{RI}(\underline{RI}(A)), \quad \forall A \in \mathcal{IF}(U),$

$$\Longleftrightarrow \text{(IUT)} \ \overline{RI}(\overline{RI}(A)) \subseteq \overline{RI}(A), \quad \forall A \in \mathcal{IF}(U).$$

(4) *R is Euclidean*
$$\Longleftrightarrow \text{(ILE)} \ \overline{RI}(\underline{RI}(A)) \subseteq \underline{RI}(A), \forall A \in \mathcal{IF}(U),$$
$$\Longleftrightarrow \text{(IUE)} \ \overline{RI}(A) \subseteq \underline{RI}(\overline{RI}(A)), \forall A \in \mathcal{IF}(U).$$

4   Basic Concepts of Intuitionistic Fuzzy Topological Spaces

In this section we introduce basic concepts related to IF topological spaces.

Definition 6. [10] *An IF topology in the sense of Lowen [27] on a nonempty set U is a family τ of IF sets in U satisfying the following axioms:*

(T$_1$) $\widehat{(\alpha, \beta)} \in \tau$ *for all* $(\alpha, \beta) \in L^*$,
(T$_2$) $G_1 \cap G_2 \in \tau$ *for any* $G_1, G_2 \in \tau$,
(T$_3$) $\bigcup\limits_{i \in J} G_i \in \tau$ *for a family* $\{G_i | i \in J\} \subseteq \tau$, *where J is an index set.*

In this case the pair (U, τ) is called an IF topological space and each IF set A in τ is referred to as an IF open set in (U, τ). The complement of an IF open set in the IF topological space (U, τ) is called an IF closed set in (U, τ).

It should be pointed out that if axiom (T$_1$) in Definition 6 is replaced by axiom
   (T$_1'$) $\emptyset \in \tau$ and $U \in \tau$,
then τ is an IF topology in the sense of Chang [6]. Clearly, an IF topology in the sense of Lowen must be an IF topology in the sense of Chang. Throughout this paper, we always consider the IF topology in the sense of Lowen.
   Now we define IF closure and interior operations in an IF topological space.

Definition 7. *Let (U, τ) be an IF topological space and $A \in \mathcal{IF}(U)$. Then the IF interior and IF closure of A are, respectively, defined as follows:*

$$int(A) = \cup\{G \mid G \text{ is an IF open set and } G \subseteq A\}, \tag{14}$$

$$cl(A) = \cap\{K \mid K \text{ is an IF closed set and } A \subseteq K\}, \tag{15}$$

and int and cl : $\mathcal{IF}(U) \to \mathcal{IF}(U)$ are, respectively, called the IF interior operator and the IF closure operator of τ, and sometimes in order to distinguish, we denote them by int_τ and cl_τ, respectively.

It can be shown that $int(A)$ is an IF open set and $cl(A)$ is an IF closed set in (U, τ), A is an IF open set in (U, τ) iff $int(A) = A$, and A is an IF closed set in (U, τ) iff $cl(A) = A$.
Moreover,

$$cl(\sim A) = \sim int(A), \quad \forall A \in \mathcal{IF}(U), \tag{16}$$

$$int(\sim A) = \sim cl(A), \quad \forall A \in \mathcal{IF}(U). \tag{17}$$

It can be verified that the IF closure operator satisfies following properties:

(Cl1) $cl(\widehat{(\alpha,\beta)}) = \widehat{(\alpha,\beta)}$, $\forall(\alpha,\beta) \in L^*$.
(Cl2) $cl(A \cup B) = cl(A) \cup cl(B)$, $\forall A, B \in \mathcal{IF}(U)$.
(Cl3) $cl(cl(A)) = cl(A)$, $\forall A \in \mathcal{IF}(U)$.
(Cl4) $A \subseteq cl(A)$, $\forall A \in \mathcal{IF}(U)$.

Properties (Cl1)–(Cl4) are called the IF closure axioms. Similarly, the IF interior operator satisfies following properties:

(Int1) $int(\widehat{(\alpha,\beta)}) = \widehat{(\alpha,\beta)}$, $\forall(\alpha,\beta) \in L^*$.
(Int2) $int(A \cap B) = int(A) \cap int(B)$, $\forall A, B \in \mathcal{IF}(U)$.
(Int3) $int(int(A)) = int(A)$, $\forall A \in \mathcal{IF}(U)$.
(Int4) $int(A) \subseteq A$, $\forall A \in \mathcal{IF}(U)$.

Definition 8. *A mapping* $cl : \mathcal{IF}(U) \to \mathcal{IF}(U)$ *is referred to as an IF closure operator if it satisfies axioms* (Cl1)–(Cl4).

Similarly, the IF interior operator can be defined by corresponding axioms.

Definition 9. *A mapping* $int : \mathcal{IF}(U) \to \mathcal{IF}(U)$ *is referred to as a fuzzy interior operator if it satisfies axioms* (Int1)-(Int4).

It is easy to show that an IF interior operator int defined in Definition 9 determines an IF topology

$$\tau_{int} = \{A \in \mathcal{IF}(U) | int(A) = A\}. \tag{18}$$

So, the IF open sets are the fixed points of int. Dually, from an IF closure operator defined in Definition 8, we can obtain an IF topology on U by setting

$$\tau_{cl} = \{A \in \mathcal{IF}(U) | cl(\sim A) =\sim A\}. \tag{19}$$

The results are summarized as the following

Theorem 7. [50] (1) *If an operator* $int : \mathcal{IF}(U) \to \mathcal{IF}(U)$ *satisfies axioms* (Int1)-(Int4), *then* τ_{int} *defined as Eq.* (18) *is an IF topology on* U *and*

$$int_{\tau_{int}} = int. \tag{20}$$

(2) *If an operator* $cl : \mathcal{IF}(U) \to \mathcal{IF}(U)$ *satisfies axioms* (Cl1)–(Cl4), *then* τ_{cl} *defined as Eq.* (19) *is an IF topology on* U *and*

$$cl_{\tau_{cl}} = cl. \tag{21}$$

Similar to the crisp Alexandrov topology [2] and crisp clopen topology [18], we now introduce the concepts of an IF Alexandrov topology and an IF clopen topology.

Definition 10. *An IF topology τ on U is called an IF Alexandrov topology if the intersection of arbitrarily many IF open sets is still open, or equivalently, the union of arbitrarily many IF closed sets is still closed. An IF topological space (U, τ) is said to be an IF Alexandrov space if τ is an IF Alexandrov topology on U. An IF topology τ on U is called an IF clopen topology if, for every $A \in \mathcal{IF}(U)$, A is IF open in (U, τ) if and only if A is IF closed in (U, τ). An IF topological space (U, τ) is said to be an IF clopen space if τ is an IF clopen topology on U.*

Theorem 8. *Let $int : \mathcal{IF}(U) \to \mathcal{IF}(U)$ be an IF interior operator. The following two conditions are equivalent:*

(1) int satisfies following property

$$\sim A \subseteq int(\sim int(A)), \forall A \in \mathcal{IF}(U). \tag{22}$$

(2) τ_{int} is an IF clopen topology, i.e., for every $A \in \mathcal{IF}(U)$, A is IF open in (U, τ) if and only if A is IF closed in (U, τ).

Proof. "(1) $\Rightarrow$ (2)." Assume that int satisfies Eq. (22). If $A \in \mathcal{IF}(U)$ is IF open, that is,

$$int(A) = A, \tag{23}$$

then

$$\sim A = \sim int(A) \subseteq int(\sim int(A)). \tag{24}$$

Since int is an IF interior operator, by (Int4), we have

$$int(\sim int(A)) \subseteq \sim int(A). \tag{25}$$

Combining Eqs. (24) and (25), we obtain

$$int(\sim int(A)) = \sim int(A). \tag{26}$$

Thus, $\sim int(A)$ is IF open. By using Eq. (23), we see that $\sim A$ is IF open, and hence A is IF closed.

On the other hand, if $A \in \mathcal{IF}(U)$ is IF closed, then $\sim A$ is IF open. By employing the above proof, we can observe that $\sim A$ is IF closed. Hence $A = \sim (\sim A)$ is IF open.

"(2) $\Rightarrow$ (1)." For any $A \in \mathcal{IF}(U)$, since $int : \mathcal{IF}(U) \to \mathcal{IF}(U)$ is an IF interior operator, $int(A)$ is open. Hence $\sim int(A)$ is IF closed. Since (U, τ_{int}) is an IF clopen space, $\sim int(A)$ is IF open. Hence

$$int(\sim int(A)) = \sim intA. \tag{27}$$

Since int is an IF interior operator, by (Int4), we have

$$int(A) \subseteq A. \tag{28}$$

Therefore, by employing Eq. (27), we conclude

$$\sim A \subseteq \sim int(A) = int(\sim int(A)). \tag{29}$$

Thus, we have proved that int satisfies Eq. (22).

5   Topological Structures of Rough Intuitionistic Fuzzy Sets

In this section we discuss the IF topological structure of rough IF sets. Throughout this section we always assume that U is a nonempty universe of discourse, R a crisp binary relation on U, and $\underline{RI}$ and $\overline{RI}$ the rough IF approximation operators defined in Definition 5.

Denote

$$\tau_R = \{A \in \mathcal{IF}(U)|\underline{RI}(A) = A\}. \tag{30}$$

The next theorem shows that any reflexive binary relation determines an IF Alexandrov topology.

Theorem 9. *If R is a reflexive crisp binary relation on U, then τ_R defined by Eq. (30) is an IF Alexandrov topology on U.*

Proof. (T$_1$) For any $(\alpha, \beta) \in L^*$, since a reflexive binary relation must be serial, in terms of Theorem 5 we have $\underline{RI}(\widetilde{(\alpha, \beta)}) = \widetilde{(\alpha, \beta)}$, thus $\widetilde{(\alpha, \beta)} \in \tau_R$.

(T$_2$) For any $A, B \in \tau_R$, that is, $\underline{RI}(A) = A$ and $\underline{RI}(B) = B$, by Theorem 4, we have $\underline{RI}(A \cap B) = \underline{RI}(A) \cap \underline{RI}(B) = A \cap B$. Thus, $A \cap B \in \tau_R$.

(T$_3$) Assume that $A_i \in \tau_R, i \in J$, J is an index set. Since R is reflexive, by (ILR) in Theorem 6, we have

$$\underline{RI}(\bigcup_{i \in J} A_i) \subseteq \bigcup_{i \in J} A_i. \tag{31}$$

For any $x \in U$, let

$$(\alpha, \beta) = (\bigcup_{i \in J} A_i)(x) = (\mu_{\underset{i \in J}{\cup} A_i}(x), \gamma_{\underset{i \in J}{\cup} A_i}(x)) = (\sup_{i \in J} \mu_{A_i}(x), \inf_{i \in J} \gamma_{A_i}(x)), \tag{32}$$

that is, $\alpha = \sup_{i \in J} \mu_{A_i}(x)$ and $\beta = \inf_{i \in J} \gamma_{A_i}(x)$. Since $\alpha = \sup_{i \in J} \mu_{A_i}(x)$, we have $\mu_{A_i}(x) \le \alpha$ for all $i \in J$ and, on the other hand, for an arbitrary $\varepsilon > 0$, there exists an $i_0 \in J$ such that $\alpha < \mu_{A_{i_0}}(x) + \varepsilon$. Since $A_i \in \tau_R$ for all $i \in J$, that is, $\underline{RI}(A_i) = A_i$ for all $i \in J$, we have $\alpha < \mu_{A_{i_0}}(x) + \varepsilon = \mu_{\underline{RI}(A_{i_0})}(x) + \varepsilon$, then, by (IL4) in Theorem 4, we conclude

$$\alpha < \mu_{\underline{RI}(A_{i_0})}(x) + \varepsilon \le \mu_{\underset{i \in J}{\cup} \underline{RI}(A_i)}(x) + \varepsilon \le \mu_{\underline{RI}(\underset{i \in J}{\cup} A_i)}(x) + \varepsilon. \tag{33}$$

By the arbitrariness of $\varepsilon > 0$, it follows that

$$\alpha \le \mu_{\underline{RI}(\underset{i \in J}{\cup} A_i)}(x), \tag{34}$$

that is,

$$\mu_{\underset{i \in J}{\cup} A_i}(x) \le \mu_{\underline{RI}(\underset{i \in J}{\cup} A_i)}(x). \tag{35}$$

Similarly, from $\beta = \inf_{i \in J} \gamma_{A_i}(x)$ we can conclude that

$$\gamma_{\bigcup_{i \in J} A_i}(x) \geq \gamma_{\underline{RI}(\bigcup_{i \in J} A_i)}(x). \tag{36}$$

Hence

$$\bigcup_{i \in J} A_i \subseteq \underline{RI}(\bigcup_{i \in J} A_i). \tag{37}$$

Combining Eqs. (31) and (37), we obtain

$$\bigcup_{i \in J} A_i = \underline{RI}(\bigcup_{i \in J} A_i). \tag{38}$$

It follows that $\bigcup_{i \in J} A_i \in \tau_R$.

Thus, τ_R is an IF topology on U. Therefore, by (IL3) and (IU3) in Theorem 4 we have proved that τ_R is an IF Alexandrov topology on U.

Theorem 10 below shows that the lower and upper rough IF approximation operators induced from a crisp reflexive and transitive relation are, respectively, an IF interior operator and an IF closure operator.

Theorem 10. *Assume that R is a crisp binary relation on U. Then the following statements are equivalent:*

(1) *R is a preorder, i.e., R is a reflexive and transitive relation;*
(2) *the upper rough IF approximation operator $\overline{RI} : \mathcal{IF}(U) \to \mathcal{IF}(U)$ is an IF closure operator;*
(3) *the lower rough IF approximation operator $\underline{RI} : \mathcal{IF}(U) \to \mathcal{IF}(U)$ is an IF interior operator.*

Proof. By the dual properties of lower and upper rough IF approximation operators, we can easily conclude that (2) and (3) are equivalent. We only need to prove that (1) and (2) are equivalent.

"(1)$\Rightarrow$(2)". Assume that R is a preorder on U. Firstly, notice that a preorder must be a serial relation, then by Theorem 5, we see that $\overline{RI}$ obeys axiom (Cl1). Secondly, according to property (IU3) in Theorem 4, we see that $\overline{RI}$ satisfies axiom (Cl2). Thirdly, since a preorder is reflexive and transitive, $\overline{RI}$ satisfies properties (IUR) and (IUT) in Theorem 6. On the other hand, properties (IUR) and (IUT) imply following property

$$\overline{RI}(A) = \overline{RI}(\overline{RI}(A)), \quad \forall A \in \mathcal{IF}(U). \tag{39}$$

Thus $\overline{RI}$ obeys axiom (Cl3). Finally, by the reflexivity of R and property (IUR) in Theorem 6, we observe that $A \subseteq \overline{RI}(A)$ for all $A \in \mathcal{IF}(U)$. Thus $\overline{RI}$ obeys axiom (Cl4). Therefore, $\overline{RI}$ is an IF closure operator.

"(2)$\Rightarrow$(1)". Assume that $\overline{RI} : \mathcal{IF}(U) \to \mathcal{IF}(U)$ is an IF closure operator. By axiom (Cl4), we see that

$$A \subseteq \overline{RI}(A), \quad \forall A \in \mathcal{IF}(U). \tag{40}$$

Then, by Theorem 6, we conclude that R is a reflexive relation. Moreover, by axiom (Cl4) again, we have

$$\overline{RI}(A) \subseteq \overline{RI}(\overline{RI}(A)), \quad \forall A \in \mathcal{IF}(U). \tag{41}$$

On the other hand, by axiom (Cl3), we observe that

$$\overline{RI}(\overline{RI}(A)) = \overline{RI}(A), \quad \forall A \in \mathcal{IF}(U). \tag{42}$$

Hence, in terms of Eqs. (41) and (42), we must have

$$\overline{RI}(\overline{RI}(A)) \subseteq \overline{RI}(A), \quad \forall A \in \mathcal{IF}(U). \tag{43}$$

According to Theorem 6, we then conclude that R is a transitive relation. Thus we have proved that R is a preorder.

Remark 2. According to Theorem 9, an IF Alexandrov topology can be obtained from a reflexive relation R by using Eq. (30), by Eq. (14) we see that any IF topology τ induces an IF interior operator int_τ, which in turn induces an IF topology τ_{int_τ}. Of course, int_τ is an IF interior operator. It also holds that $\tau_{int_\tau} = \tau$. Now let us take τ_R, then its interior is int_{τ_R} which produces an IF topology $\tau_{int_{\tau_R}}$. Since $\tau_{int_{\tau_R}} = \tau_R$, we have:

$$\tau_R = \{A \in \mathcal{IF}(U) | \underline{RI}(A) = A\} = \{A \in \mathcal{IF}(U) | int_{\tau_R}(A) = A\}. \tag{44}$$

We should point out that though $\underline{RI}$ and int_{τ_R} produce the same IF topology, in general, $\underline{RI} \neq int_{\tau_R}$, the reason is that int_{τ_R} is an IF interior operator whereas $\underline{RI}$ is not. In fact, Theorem 10 shows that $\underline{RI} = int_{\tau_R}$ if and only if R is a preorder.

Lemma 1. *If R is a symmetric crisp binary relation on U, then for all $A, B \in \mathcal{IF}(U)$,*

$$\overline{RI}(A) \subseteq B \Longleftrightarrow A \subseteq \underline{RI}(B). \tag{45}$$

Proof. "$\Rightarrow$" For $A, B \in \mathcal{IF}(U)$, if $\overline{RI}(A) \subseteq B$, by (IL1) and (IU1) in Theorem 4, we have $\sim \underline{RI}(\sim A) \subseteq B$, then, $\sim B \subseteq \underline{RI}(\sim A)$. By (IU5) in Theorem 4 and (ILS) in Theorem 6, it follows that

$$\overline{RI}(\sim B) \subseteq \overline{RI}(\underline{RI}(\sim A)) \subseteq \sim A. \tag{46}$$

Hence

$$A \subseteq \sim \overline{RI}(\sim B) = \underline{RI}(B). \tag{47}$$

"$\Leftarrow$" For $A, B \in \mathcal{IF}(U)$, assume that $A \subseteq \underline{RI}(B)$, by (IL1) in Theorem 4, we have $A \subseteq \sim \overline{RI}(\sim B)$, that is, $\overline{RI}(\sim B) \subseteq \sim A$, according to (IL5) in Theorem 4, we then conclude

$$\underline{RI}(\overline{RI}(\sim B)) \subseteq \underline{RI}(\sim A). \tag{48}$$

Furthermore, by (IL1) and (IU1) in Theorem 4, it is easy to obtain that

$$\sim \overline{RI}(\underline{RI}(B)) \subseteq \underline{RI}(\sim A) = \sim \overline{RI}(A). \tag{49}$$

Consequently, by (ILS) in Theorem 6, we conclude that

$$\overline{RI}(A) \subseteq \overline{RI}(\underline{RI}(B)) \subseteq B. \tag{50}$$

Theorem 11. *Let R be a similarity crisp binary relation on U, and $\underline{RI}$ and $\overline{RI}$ the rough IF approximation operators defined in Definition 5. Then $\underline{RI}$ and $\overline{RI}$ satisfy property (Clop): for $A \in \mathcal{IF}(U)$,*

$$\underline{RI}(A) = A \Longleftrightarrow A = \overline{RI}(A) \Longleftrightarrow \underline{RI}(\sim A) = \sim A \Longleftrightarrow \sim A = \overline{RI}(\sim A). \quad (51)$$

Proof. For $A \in \mathcal{IF}(U)$, assume that $\underline{RI}(A) = A$. Since R is reflexive, by (ILR) in Theorem 6, we see that $\underline{RI}(A) = A$ implies $A \subseteq \underline{IR}(A)$. Then, by Lemma 1, we conclude $\overline{RI}(A) \subseteq A$. Furthermore, since R is reflexive, in terms of (IUR) in Theorem 6 we obtain $A = \overline{RI}(A)$. Similarly, we can prove that $A = \overline{RI}(A)$ implies $\underline{RI}(A) = A$. Moreover, by using (IL1) and (IU1) in Theorem 4, it is easy to verify that Eq. (51) holds.

The next theorem shows that an IF topological space induced from a reflexive and symmetric crisp approximation space is an IF clopen topological space.

Theorem 12. *Let R be a similarity crisp binary relation on U, and $\underline{RI}$ and $\overline{RI}$ the rough IF approximation operators defined in Definition 5. Then τ_R defined as Eq. (30) is an IF clopen topology on U.*

Proof. For $A \in \mathcal{IF}(U)$, since R is a similarity crisp binary relation, by Theorem 11, we have

$$
\begin{aligned}
A \text{ is IF open} &\Longleftrightarrow A \in \tau_R \\
&\Longleftrightarrow A = \underline{RI}(A) \\
&\Longleftrightarrow \sim A = \underline{RI}(\sim A) \\
&\Longleftrightarrow \sim A \in \tau_R \\
&\Longleftrightarrow A \text{ is IF closed.}
\end{aligned}
$$

Thus, τ_R is an IF clopen topology on U.

Corollary 1. *Let R be an equivalent crisp binary relation on U, and $\underline{RI}$ and $\overline{RI} : \mathcal{IF}(U) \to \mathcal{IF}(U)$ the rough IF approximation operators of (U, R). Let $\tau_R = \{A \in \mathcal{IF}(U) | \underline{RI}(A) = A\}$, then*

(1) $\underline{RI}$ and $\overline{RI} : \mathcal{IF}(U) \to \mathcal{IF}(U)$ are, respectively, the IF interior operator and the IF closure operator of the IF topology τ_R
(2) τ_R is an IF clopen topology on U.

6 The Conditions of an Intuitionistic Fuzzy Topological Spaces Induced from a Crisp Approximation Space

As can be seen from Sect. 5, a crisp preorder can produce an IF topological space such that its IF interior and closure operators are, respectively, the lower and upper rough IF approximation operators of the given crisp approximation space. In this section, we consider the reverse problem, that is, under which conditions can an IF topological space be associated with a crisp approximation space which produces the given IF topological space?

Definition 11. *Let* $O : \mathcal{IF}(U) \to \mathcal{IF}(U)$ *be an IF operator, we define two operators from* $\mathcal{F}(U)$ *to* $\mathcal{F}(U)$, *denoted* O_μ *and* O_γ *respectively, such that for* $A \in \mathcal{IF}(U)$,

$$O_\mu(\mu_A) = \mu_{O(A)} \text{ and } O_\gamma(\gamma_A) = \gamma_{O(A)}. \tag{52}$$

That is,

$$O(A) = O((\mu_A, \gamma_A)) = (\mu_{O(A)}, \gamma_{O(A)}) = (O_\mu(\mu_A), O_\gamma(\gamma_A)). \tag{53}$$

By Eqs. (12) and (13), we can observe that for $A \in \mathcal{IF}(U)$,

$$\underline{RI}_\mu(\mu_A) = \underline{RF}(\mu_A), \underline{RI}_\gamma(\gamma_A) = \overline{RF}(\gamma_A). \tag{54}$$

$$\overline{RI}_\mu(\mu_A) = \overline{RF}(\mu_A), \overline{RI}_\gamma(\gamma_A) = \underline{RF}(\gamma_A). \tag{55}$$

The following Theorem 13 gives the sufficient and necessary conditions that an IF interior (closure, respectively) operator in an IF topological space can be associated with a crisp reflexive and transitive relation such that the induced lower (upper, respectively) rough IF approximation operator is exactly the given IF interior (closure, respectively) operator.

Theorem 13. *Let* (U, τ) *be an IF topological space and* $cl, int : \mathcal{IF}(U) \to \mathcal{IF}(U)$ *its IF closure operator and IF interior operator, respectively. Then there exists a crisp reflexive and transitive relation* R *on* U *such that*

$$\overline{RI}(A) = cl(A) \text{ and } \underline{RI}(A) = int(A), \ \forall A \in \mathcal{IF}(U) \tag{56}$$

iff cl *satisfies the following conditions* (C1)-(C3), *or equivalently,* int *obeys the following conditions* (I1)-(I3):

(C1) $cl(1_y) \in \mathcal{P}(U), \forall y \in U$,
(C2) $cl(A \cap \widehat{(\alpha, \beta)}) = cl(A) \cap \widehat{(\alpha, \beta)}, \ \forall A \in \mathcal{IF}(U), \ \forall (\alpha, \beta) \in L^*$.
(C3) $cl(\bigcup_{i \in J} A_i) = \bigcup_{i \in J} cl(A_i), \ \forall A_i \in \mathcal{IF}(U), \ \forall i \in J, \ J \text{ is any index set.}$
(I1) $int(1_{U - \{y\}}) \in \mathcal{P}(U), \forall y \in U$,
(I2) $int(A \cup \widehat{(\alpha, \beta)}) = int(A) \cup \widehat{(\alpha, \beta)}, \ \forall A \in \mathcal{IF}(U), \ \forall (\alpha, \beta) \in L^*$.
(I3) $int(\bigcap_{i \in J} A_i) = \bigcap_{i \in J} int(A_i), \ \forall A_i \in \mathcal{IF}(U), \ \forall i \in J, \ J \text{ is any index set.}$

Proof. "$\Rightarrow$" Assume that there exists a crisp reflexive and transitive relation R_τ on U such that Eq. (56) holds, then, by Theorem 4, it can be seen that conditions (C1)-(C3), and (I1)-(I3) hold.

"$\Leftarrow$" Assume that the closure operator $cl : \mathcal{IF}(U) \to \mathcal{IF}(U)$ satisfies conditions (C1)-(C3) and the interior operator $int : \mathcal{IF}(U) \to \mathcal{IF}(U)$ obeys conditions (I1)-(I3). For the closure operator $cl : \mathcal{IF}(U) \to \mathcal{IF}(U)$, by Definition 11, we can derive two operators cl_μ and cl_γ from $\mathcal{F}(U)$ to $\mathcal{F}(U)$ such that, for all $A \in \mathcal{IF}(U)$,

$$cl_\mu(\mu_A) = \mu_{cl(A)} \text{ and } cl_\gamma(\gamma_A) = \gamma_{cl(A)}. \tag{57}$$

Likewise, from the interior operator int we can obtain two operators int_μ and int_γ from $\mathcal{F}(U)$ to $\mathcal{F}(U)$ such that, for all $A \in \mathcal{IF}(U)$,

$$int_\mu(\mu_A) = \mu_{int(A)} \text{ and } int_\gamma(\gamma_A) = \gamma_{int(A)}. \tag{58}$$

For $y \in U$, notice that $cl(1_y) \in \mathcal{P}(U)$ if and only if $\mu_{cl(1_y)}(x) \in \{0,1\}$ and $\mu_{cl(1_y)}(x) + \gamma_{cl(1_y)}(x) = 1$ for all $x \in U$. Then we can define a crisp binary relation R on U by employing cl as follows: for $(x,y) \in U \times U$,

$$(x,y) \in R \Longleftrightarrow \mu_{cl(1_y)}(x) = 1, \tag{59}$$

that is,

$$y \in R_s(x) \Longleftrightarrow cl_\mu(\mu_{1_y})(x) = 1 \Longleftrightarrow cl_\gamma(\gamma_{1_y})(x) = 0. \tag{60}$$

Note that, for $A \in \mathcal{IF}(U)$,

$$\mu_A = \bigcup_{y \in U} [\mu_{1_y} \cap \widehat{\mu_A(y)}], \tag{61}$$

$$\gamma_A = \bigcap_{y \in U} [\gamma_{1_y} \cup \widehat{\gamma_A(y)}]. \tag{62}$$

On the other hand, it can be verified that (C2) implies (CF2) and (CF2)′, and (C3) implies (CF3) and (CF3)′:

(CF2) $cl_\mu(\mu_{A \cap \widehat{(\alpha,\beta)}}) = cl_\mu(\mu_A \cap \widehat{\alpha}) = cl_\mu(\mu_A) \cap \widehat{\alpha}$, $\forall A \in \mathcal{IF}(U), \forall(\alpha,\beta) \in L^*$.

(CF2)′ $cl_\gamma(\gamma_{A \cap \widehat{(\alpha,\beta)}}) = cl_\gamma(\gamma_A \cup \widehat{\beta}) = cl_\gamma(\gamma_A) \cup \widehat{\beta}$, $\forall A \in \mathcal{IF}(U), \forall(\alpha,\beta) \in L^*$.

(CF3) $cl_\mu(\mu_{\bigcup_{i \in J} A_i}) = cl_\mu(\bigcup_{i \in J} \mu_{A_i}) = \bigcup_{i \in J} cl_\mu(\mu_{A_i})$, $\forall A_i \in \mathcal{IF}(U), i \in J, J$ is any index set.

(CF3)′ $cl_\gamma(\gamma_{\bigcup_{i \in J} A_i}) = cl_\gamma(\bigcap_{i \in J} \gamma_{A_i}) = \bigcap_{i \in J} cl_\gamma(\gamma_{A_i})$, $\forall A_i \in \mathcal{IF}(U), i \in J, J$ is any index set.

Then, for any $x \in U$, in terms of Definition 11 and above properties, we have

$$\begin{aligned}
\mu_{\overline{RI(A)}}(x) &= \bigvee_{y \in R_s(x)} \mu_A(y) \\
&= \bigvee_{y \in U} [1_{R_s(x)}(y) \wedge \mu_A(y)] \\
&= \bigvee_{y \in U} [cl_\mu(\mu_{1_y})(x) \wedge \mu_A(y)] \\
&= \bigvee_{y \in U} [(cl_\mu(\mu_{1_y}) \cap \widehat{\mu_A(y)})(x)] \\
&= \bigvee_{y \in U} [(cl_\mu(\mu_{1_y} \cap \widehat{\mu_A(y)}))(x)] \\
&= [\bigcup_{y \in U} (cl_\mu(\mu_{1_y} \cap \widehat{\mu_A(y)}))](x) \\
&= [cl_\mu(\bigcup_{y \in U} (\mu_{1_y} \cap \widehat{\mu_A(y)}))](x) \\
&= cl_\mu(\mu_A)(x) = \mu_{cl(A)}(x).
\end{aligned}$$

And likewise, we obtain

$$
\begin{aligned}
\gamma_{\overline{RI}(A)}(x) &= \bigwedge_{y \in R_s(x)} \gamma_A(y) \\
&= \bigwedge_{y \in U} \left[(1 - 1_{R_s(x)}(y)) \vee \gamma_A(y) \right] \\
&= \bigwedge_{y \in U} \left[cl_\gamma(\gamma_{1_y})(x) \vee \gamma_A(y) \right] \\
&= \bigwedge_{y \in U} \left[(cl_\gamma(\gamma_{1_y}) \cup \widehat{\gamma_A(y)})(x) \right] \\
&= \bigwedge_{y \in U} \left[(cl_\gamma(\gamma_{1_y} \cup \widehat{\gamma_A(y)}))(x) \right] \\
&= \left[\bigcap_{y \in U} (cl_\gamma(\gamma_{1_y} \cup \widehat{\gamma_A(y)})) \right](x) \\
&= \left[cl_\gamma(\bigcap_{y \in U} (\gamma_{1_y} \cup \widehat{\gamma_A(y)})) \right](x) \\
&= cl_\gamma(\gamma_A)(x) = \gamma_{cl(A)}(x).
\end{aligned}
$$

Therefore, $cl(A) = \overline{RI}(A)$.

Similarly, we can conclude that $int(A) = \underline{RI}(A)$. Furthermore, by Theorem 10, we can observe that the binary relation R is reflexive and transitive.

Let $\mathcal{R}$ be the set of all crisp preorders on U and $\mathcal{T}$ the set of all IF Alexandrov spaces on U in which the IF interior operator satisfies axioms (I1)-(I3), and the IF closure operator obeys axioms (C1)-(C3). Then, we can easily conclude following Theorems 14 and 15.

Theorem 14. (1) If $R \in \mathcal{R}$, τ_R is defined by Eq. (30) and R_{τ_R} is defined by Eq. (59), then $R_{\tau_R} = R$.

(2) If $\tau \in \mathcal{T}$, R_τ is defined by Eq. (59), and τ_{R_τ} is defined by Eq. (30), then $\tau_{R_\tau} = \tau$.

Theorem 15. *There exists a one-to-one correspondence between $\mathcal{R}$ and $\mathcal{T}$.*

Theorem 16. *Let (U, τ) be an IF topological space and $cl_\tau, int_\tau : \mathcal{IF}(U) \to \mathcal{IF}(U)$ its IF closure and interior operators, respectively. If there exists a crisp binary R_τ on U such that Eq. (56) holds, then (U, τ) is an IF clopen space if and only R_τ is an equivalence relation on U.*

Proof. "⇒" If exists a crisp binary R on U such that Eq. (56) holds, then, by Theorem 10, we conclude that R is a preorder. On the other hand, since (U, τ) is an IF clopen topological space, according to Theorem 8, we see that the IF interior operator $int_\tau : \mathcal{IF}(U) \to \mathcal{IF}(U)$ satisfies Eq. (22). Consequently, it is easy to verify that $\underline{RI}, \overline{RI} : \mathcal{IF}(U) \to \mathcal{IF}(U)$ obey properties (ILS). Hence, by Theorem 6, we conclude that R is symmetric. Therefore, R is an equivalence relation on U.

"⇐" It follows immediately from Corollary 1.

7 Conclusion

In this paper we have studied the topological structure of rough IF sets. We have shown that a reflexive crisp relation can induce an IF Alexandrov space. We have also examined that the lower and upper rough IF approximation operators are, respectively, an IF interior operator and an IF closure operator if and only if the binary relation in the crisp approximation space is reflexive and transitive. We have further presented that an IF topological space induced from a reflexive and symmetric crisp approximation space is an IF clopen topological space. Finally, we have explored the sufficient and necessary conditions that an IF interior (closure, respectively) operator derived from an IF topological space can associate with a reflexive and transitive crisp approximation space such that the induced lower (upper, respectively) rough IF approximation operator is exactly the IF interior (closure, respectively) operator.

Acknowledgments. This work was supported by grants from the National Natural Science Foundation of China (Nos. 61272021, 61075120, 11071284, 61202206, and 61173181), the Zhejiang Provincial Natural Science Foundation of China (No. LZ12F03002), and Chongqing Key Laboratory of Computational Intelligence (No. CQ-LCI-2013-01).

References

1. Albizuri, F.X., Danjou, A., Grana, M., Torrealdea, J., Hernandezet, M.C.: The high-order Boltzmann machine: learned distribution and topology. IEEE Trans. Neural Netw. **6**, 767–770 (1995)
2. Arenas, F.G.: Alexandroff Spaces. Acta Math. Univ. Comenianae. **68**, 17–25 (1999)
3. Atanassov, K.: Intuitionistic Fuzzy Sets. Physica-Verlag, Heidelberg (1999)
4. Boixader, D., Jacas, J., Recasens, J.: Upper and lower approximations of fuzzy sets. Int. J. Gen. Syst. **29**, 555–568 (2000)
5. Chakrabarty, K., Gedeon, T., Koczy, L.: Intuitionistic fuzzy rough set. In: Proceedings of 4th Joint Conference on Information Sciences (JCIS), Durham, NC, pp. 211–214 (1998)
6. Chang, C.L.: Fuzzy topological spaces. J. Math. Anal. Appl. **24**, 182–189 (1968)
7. Choudhury, M.A., Zaman, S.I.: Learning sets and topologies. Kybernetes. **35**, 1567–1578 (2006)
8. Chuchro, M.: On rough sets in topological boolean algebras. In: Ziarko, W. (ed.) Rough Sets, Fuzzy Sets and Knowledge Discovery, pp. 157–160. Springer, Berlin (1994)
9. Chuchro, M.: A certain conception of rough sets in topological boolean algebras. Bull. Sect. Logic. **22**, 9–12 (1993)
10. Coker, D.: An introduction of intuitionistic fuzzy topological spaces. Fuzzy Sets Syst. **88**, 81–89 (1997)
11. Coker, D.: Fuzzy rough sets are intuitionistic *L*-fuzzy sets. Fuzzy Sets Syst. **96**, 381–383 (1998)
12. Cornelis, C., Cock, M.D., Kerre, E.E.: Intuitionistic fuzzy rough sets: at the crossroads of imperfect knowledge. Expert Syst. **20**, 260–270 (2003)

13. Cornelis, C., Deschrijver, G., Kerre, E.E.: Implication in Intuitionistic fuzzy and interval-valued fuzzy set theory: construction, classification application. Int. J. Approximate Reasoning **35**, 55–95 (2004)
14. Dubois, D., Prade, H.: Rough fuzzy sets and fuzzy rough sets. Int. J. Gen. Syst. **17**, 191–209 (1990)
15. Hao, J., Li, Q.G.: The relationship between L-Fuzzy rough set and L-Topology. Fuzzy Sets Syst. **178**, 74–83 (2011)
16. Jena, S.P., Ghosh, S.K.: Intuitionistic fuzzy rough sets. Notes Intuitionistic Fuzzy Sets **8**, 1–18 (2002)
17. Kall, L., Krogh, A., Sonnhammer, E.L.L.: A combined transmembrane topology and signal peptide prediction method. J. Mol. Biol. **338**, 1027–1036 (2004)
18. Kondo, M.: On the structure of generalized rough sets. Inf. Sci. **176**, 589–600 (2006)
19. Kortelainen, J.: On relationship between modified sets, topological space and rough sets. Fuzzy Sets Syst. **61**, 91–95 (1994)
20. Kortelainen, J.: On the evaluation of compatibility with gradual rules in information systems: a topological approach. Control Cybern. **28**, 121–131 (1999)
21. Kortelainen, J.: Applying modifiers to knowledge acquisition. Inf. Sci. **134**, 39–51 (2001)
22. Lai, H.L., Zhang, D.X.: Fuzzy preorder and fuzzy topology. Fuzzy Sets Syst. **157**, 1865–1885 (2006)
23. Lashin, E.F., Kozae, A.M., Khadra, A.A.A., Medhat, T.: Rough set theory for topological spaces. Int. J. Approximate Reasoning **40**, 35–43 (2005)
24. Li, J.J., Zhang, Y.L.: Reduction of subbases and its applications. Utilitas Math. **82**, 179–192 (2010)
25. Li, Y.L., Li, Z.L., Chen, Y.Q., Li, X.X.: Using raster quasi-topology as a tool to study spatial entities. Kybernetes **32**, 1425–1449 (2003)
26. Li, Z.W., Xie, T.S., Li, Q.G.: Topological structure of generalized rough sets. Comput. Math. Appl. **63**, 1066–1071 (2012)
27. Lowen, R.: Fuzzy topological spaces and fuzzy compactness. J. Math. Anal. Appl. **56**, 621–633 (1976)
28. Mi, J.-S., Leung, Y., Zhao, H.-Y., Feng, T.: Generalized fuzzy rough sets determined by a triangular norm. Inform. Sci. **178**, 3203–3213 (2008)
29. Mi, J.-S., Zhang, W.-X.: An axiomatic characterization of a fuzzy generalization of rough sets. Inf. Sci. **160**, 235–249 (2004)
30. Morsi, N.N., Yakout, M.M.: Axiomatics for fuzzy rough sets. Fuzzy Sets Syst. **100**, 327–342 (1998)
31. Pawlak, Z.: Rough Sets: Theoretical Aspects of Reasoning about Data. Kluwer Academic Publishers, Boston (1991)
32. Pei, Z., Pei, D.W., Zheng, L.: Topology vs generalized rough sets. Int. J. Approximate Reasoning **52**, 231–239 (2011)
33. Qin, K.Y., Pei, Z.: On the topological properties of fuzzy rough sets. Fuzzy Sets Syst. **151**, 601–613 (2005)
34. Qin, K.Y., Yang, J., Pei, Z.: Generalized rough sets based on reflexive and transitive relations. Inf. Sci. **178**, 4138–4141 (2008)
35. Radzikowska, A.M., Kerre, E.E.: A comparative study of fuzzy rough sets. Fuzzy Sets Syst. **126**, 137–155 (2002)
36. Rizvi, S., Naqvi, H.J., Nadeem, D.: Rough intuitionistic fuzzy set. In: Proceedings of the 6th Joint Conference on Information Sciences (JCIS), Durham, NC, pp. 101–104 (2002)

37. Samanta, S.K., Mondal, T.K.: Intuitionistic fuzzy rough sets and rough intuitionistic fuzzy sets. J. Fuzzy Math. **9**, 561–582 (2001)
38. Thiele, H.: On axiomatic characterisation of fuzzy approximation operators II, the rough fuzzy set based case. In: Proceedings of the 31st IEEE International Symposium on Multiple-Valued Logic, pp. 330–335 (2001)
39. Tiwari, S.P., Srivastava, A.K.: Fuzzy rough sets, fuzzy preorders and fuzzy topologies. Fuzzy Sets Syst. **210**, 63–68 (2013)
40. Wiweger, R.: On topological rough sets. Bull. Pol. Acad. Sci. Math. **37**, 89–93 (1989)
41. Wu, Q.E., Wang, T., Huang, Y.X., Li, J.S.: Topology theory on rough sets. IEEE Trans. Syst. Man Cybern. Part B-Cybern **38**, 68–77 (2008)
42. Wu, W.-Z.: A study on relationship between fuzzy rough approximation operators and fuzzy topological spaces. In: Wang, L., Jin, Y. (eds.) FSKD 2005. LNCS (LNAI), vol. 3613, pp. 167–174. Springer, Heidelberg (2005)
43. Wu, W.-Z.: On some mathematical structures of T-fuzzy rough set algebras in infinite universes of discourse. Fundam. Informatica **108**, 337–369 (2011)
44. Wu, W.-Z., Leung, Y., Mi, J.-S.: On characterizations of $(\mathcal{I}, \mathcal{T})$-fuzzy rough approximation operators. Fuzzy Sets Syst. **154**, 76–102 (2005)
45. Wu, W.-Z., Leung, Y., Shao, M.-W.: Generalized fuzzy rough approximation operators determined by fuzzy implicators. Int. J. Approximate Reasoning **54**, 1388–1409 (2013)
46. Wu, W.-Z., Leung, Y., Zhang, W.-X.: On generalized rough fuzzy approximation operators. In: Peters, J.F., Skowron, A. (eds.) Transactions on Rough Sets V. LNCS, vol. 4100, pp. 263–284. Springer, Heidelberg (2006)
47. Wu, W.-Z., Xu, Y.-H.: Rough approximations of intuitionistic fuzzy sets in crisp approximation spaces. In: Proceedings of Seventh International Conference on Fuzzy Systems and Knowledge Discovery (FSKD 2010), vol. 1, pp. 309–313 (2010)
48. Wu, W.-Z., Xu, Y.-H.: On Fuzzy Topological Structures of Rough Fuzzy Sets. In: Peters, J.F., Skowron, A., Ramanna, S., Suraj, Z., Wang, X. (eds.) Transactions on Rough Sets XVI. LNCS, vol. 7736, pp. 125–143. Springer, Heidelberg (2013)
49. Wu, W.-Z., Zhang, W.-X.: Constructive and axiomatic approaches of fuzzy approximation operators. Inf. Sci. **159**, 233–254 (2004)
50. Wu, W.-Z., Zhou, L.: On intuitionistic fuzzy topologies based on intuitionistic fuzzy reflexive and transitive relations. Soft Comput. **15**, 1183–1194 (2011)
51. Yang, L.Y., Xu, L.S.: Topological properties of generalized approximation spaces. Inf. Sci. **181**, 3570–3580 (2011)
52. Yao, Y.Y.: Constructive and algebraic methods of the theory of rough sets. J. Inf. Sci. **109**, 21–47 (1998)
53. Zhang, H.-P., Yao, O.Y., Wang, Z.D.: Note on "Generlaized Rough Sets Based on Reflexive and Transitive Relations". Inf. Sci. **179**, 471–473 (2009)
54. Zhou, L., Wu, W.-Z.: On generalized intuitionistic fuzzy approximation operators. Inf. Sci. **178**, 2448–2465 (2008)
55. Zhou, L., Wu, W.-Z., Zhang, W.-X.: On intuitionistic fuzzy rough sets and their topological structures. Int. J. Gen. Syst. **38**, 589–616 (2009)
56. Zhou, L., Wu, W.-Z., Zhang, W.-X.: On characterization of intuitionistic fuzzy rough sets based on intuitionistic fuzzy implicators. Inf. Sci. **179**, 883–898 (2009)
57. Zhu, W.: Topological approaches to covering rough sets. Inf. Sci. **177**, 1499–1508 (2007)

Feature Selection with Positive Region Constraint for Test-Cost-Sensitive Data

Jiabin Liu[1,2], Fan Min[2(✉)], Hong Zhao[2], and William Zhu[2]

[1] Department of Computer Science, Sichuan University for Nationalities,
Kangding 626001, China
liujb418@163.com
[2] Lab of Granular Computing, Zhangzhou Normal University,
Zhangzhou 363000, China
minfanphd@163.com

Abstract. In many data mining and machine learning applications, data are not free, and there is a test cost for each data item. Due to economic, technological and legal reasons, it is neither possible nor necessary to obtain a classifier with 100 % accuracy. In this paper, we consider such a situation and propose a new constraint satisfaction problem to address it. With this in mind, one has to minimize the test cost to keep the accuracy of the classification under a budget. The constraint is expressed by the positive region, whereas the object is to minimizing the total test cost. The new problem is essentially a dual of the test cost constraint attribute reduction problem, which has been addressed recently. We propose a heuristic algorithm based on the information gain, the test cost, and a user specified parameter λ to deal with the new problem. The algorithm is tested on four University of California - Irvine datasets with various test cost settings. Experimental results indicate that the algorithm finds optimal feature subset in most cases, the rational setting of λ is different among datasets, and the algorithm is especially stable when the test cost is subject to the Pareto distribution.

Keywords: Feature selection · Cost-sensitive learning · Positive region · Test cost

1 Introduction

When industrial products are manufactured, they must be inspected strictly before delivery. Testing equipments are needed to classify the product as qualified, unqualified, etc. Each equipment costs money, which will be averaged on each product. Generally, we should pay more to obtain higher classification accuracy. However, due to economic, technological and legal reasons, it is neither possible nor necessary to obtain a classifier with 100 % accuracy. There may be an industrial standard to indicate the accuracy of the classification, such as 95 %. Consequently, we are interested in a set of equipments with minimal cost meeting the standard. In this scenario, there are two issues: one is the equipment

© Springer-Verlag Berlin Heidelberg 2014
J.F. Peters et al. (Eds.): Transactions on Rough Sets XVIII, LNCS 8449, pp. 23–33, 2014.
DOI: 10.1007/978-3-662-44680-5_2

cost, and the other is the product classification accuracy. They are called test cost and classification accuracy, respectively. Since the classification accuracy only needs to meet the industrial standard, we would like to choose some testing equipments to minimize the total cost.

In many real applications of data mining, machine learning, pattern recognition and signal processing, datasets often contain huge numbers of features. In such a case, feature selection will be necessary [15]. Feature selection [10,32,42] is the process of choosing a subset of features from the original set of features forming patterns in a given dataset. The subset should be necessary and sufficient to describe target concepts, retaining a suitably high accuracy in representing the original features. Feature selection serves as a pre-processing technique in machine learning and pattern recognition application [1]. Consequently, it has been defined by many authors by booking at it from various angles [2].

According to Pawlak [27], minimal reducts have the best generalization ability. Hence many feature selection reduction algorithms based on rough set have been proposed to find one of them (see, e.g., [29,31,36,38,46]). On the other side, however, data are not free, and there is a test cost for each data item [9]. Therefore the classifier should also exhibit low test cost [19]. With this in mind, the minimal test cost reduct problem has been defined [17] and with some algorithms proposed to deal with it [6,7,21,25,35,41].

In this paper, we formally define the feature selection with positive region constraint for test-cost-sensitive data (FSPRC) problem. The positive region is a widely used concept in rough set [26]. We use this concept instead of the classification accuracy to specify the industrial standard. The new problem is essentially a dual of the optimal sub-reduct with test cost constraint (OSRT) problem, which has been defined in [20] and studied in [14,22,23]. The OSRT problem considers the test cost constraint, while the new problem considers the positive region constraint. As will be discussed in the following text, the classical reduct problem can be viewed as a special case of the FSPRC problem. Since the classical reduct problem is NP-hard, the new problem is at least NP-hard.

We propose a heuristic algorithm to deal with the new problem. The algorithm follows a popular addition-deletion framework. Since we do not require that the positive region is unchanged after feature selection, there does not exist a core computing stage. The heuristic information function is based on both the information gain and the test cost. It is deliberately designed to obtain a tradeoff between the usefulness and the cost of each feature.

Four open datasets from the University of California-Irvine (UCI) library are employed to study the performance and effectiveness of our algorithms. Experiments are undertaken with open source software Cost-sensitive rough sets (Coser) [24]. Results indicate the algorithm can find the optimal feature subset in most cases, the rational setting of λ is different among datasets, and the algorithm is especially stable when the test cost is subject to the Pareto distribution.

The rest of this paper is organized as follows. Section 2 describes related concepts in the rough set theory and defines the FSPRC problem formally. In Sect. 3, a heuristic algorithm based on λ-weighted information gain is presented.

Section 4 illustrates some results on four UCI datasets with detailed analysis. Finally, Sect. 5 gives some conclusions and indicates possibilities for further work.

2   Preliminaries

In this section, we define the FSPRC problem. First, we revisit the data model on which the problem is defined. Then we review the concept of positive region. Finally we propose feature selection with positive region constraint problem.

2.1   Test-Cost-Independent Decision Systems

Decision systems are fundamental in machine learning and data mining. A decision system is often denoted as $S = (U, C, D, \{V_a | a \in C \cup D\}, \{I_a | a \in C \cup D\})$, where U is a finite set of objects called the universe, C is the set of conditional attributes, D is the set of decision attributes, V_a is the set of values for each $a \in C \cup D$, and $I_a : U \to V_a$ is an information function for each $a \in C \cup D$. We often denote $\{V_a | a \in C \cup D\}$ and $\{I_a | a \in C \cup D\}$ by V and I, respectively. Table 1 is a decision system, which conditional attributes are symbolic values. Here $C = \{$Patient, Headache, Temperature, Lymphocyte, Leukocyte, Eosinophil, Heartbeat$\}$, $d = \{$Flu$\}$, and $U = \{x_1, x_2, \ldots, x_7\}$.

A *test-cost-independent decision system* (TCI-DS) [19] is a decision system with test cost information represented by a vector, as the one in Table 2. It is the simplest form of the test-cost-sensitive decision system and defined as follows.

Table 1. An exemplary decision table

Patient	Headache	Temperature	Lymphocyte	Leukocyte	Eosinophil	Heartbeat	Flu
x_1	Yes	High	High	High	High	Normal	Yes
x_2	Yes	High	Normal	High	High	Abnormal	Yes
x_3	Yes	High	High	High	Normal	Abnormal	Yes
x_4	No	High	Normal	Normal	Normal	Normal	No
x_5	Yes	Normal	Normal	Low	High	Abnormal	No
x_6	Yes	Normal	Low	High	Normal	Abnormal	No
x_7	Yes	Low	Low	High	Normal	Normal	Yes

Table 2. An exemplary cost vector

a	Headache	Temperature	Lymphocyte	Leukocyte	Eosinophil	Heartbeat
$c(a)$	\$12	\$5	\$15	\$20	\$15	\$10

Definition 1. *[19] A test-cost-independent decision system (TCI-DS) S is the 6-tuple:*

$$S = (U, C, D, V, I, c), \tag{1}$$

where U, C, D, V and I have the same meanings as in a decision system, and $c : C \rightarrow R^+ \cup \{0\}$ is the test cost function. Here the test cost function can easily be represented by a vector $c = [c(a_1), c(a_2), \ldots, c(a_{|C|})]$, which indicates that test costs are independent of one another, that is, $c(B) = \sum_{a \in B} c(a)$ for any $B \subset C$.

For example, if we select tests Temperature, Lymphocyte, Leukocyte and Heartbeat, the total test cost would be \$5 + \$15 + \$20 + \$10 = \$50. This is why we call this type of decision system "test-cost-independent". If any element in c is 0, a TCI-DS coincides with a DS. Therefore, free tests are not considered for the sake of simplicity. A TCI-DS is represented by a decision table and a test cost vector. One example is given by Tables 1 and 2 [19].

2.2 Positive Region

Rough set theory [27] is an approach to vagueness and uncertainty. Similarly to fuzzy set theory it is not an alternative to classical set theory but it is embedded in it. Positive region is an important concept in rough set theory. It is defined by through lower approximation. Let $S = (U, C, D, V, I)$ be a decision system. Any $\emptyset \neq B \subseteq C \cup D$ determines an indiscernibility relation on U. A partition determined by B is denoted by U/B. Let $\underline{B}(X)$ denote the B-lower approximation of X.

Definition 2. *[27] Let $S = (U, C, D, V, I)$ be a decision system, $\forall B \subset C$, the positive region of D with respect to B is defined as*

$$POS_B(D) = \bigcup_{X \in U/D} \underline{B}(X), \tag{2}$$

where U, C, D, V and I have the same meanings as in a decision system.

In other words, D is totally (partially) dependent on B, if all (some) elements of the universe U can be uniquely classified to blocks of the partition U/D, employing B [26].

2.3 Feature Selection with Positive Region Constraint Problem

Feature selection is the process of choosing an appropriate subset of attributes from the original dataset [34]. There are numerous reduct problems which have been defined on the classical [28], the neighborhood (see, e.g., [11,12]), the covering-based [16,40,43–46], the decision-theoretical [37], and the dominance-based [4] rough set models. Respective definitions of relative reducts also have been studied in [8,29].

Definition 3. *[27] Let $S = (U, C, D, V, I)$ be a decision system. Any $B \subseteq C$ is called a decision relative reduct (or a relative reduct for brevity) of S iff:*

(1) $POS_B(D) = POS_C(D)$.

(2) $\forall a \in B, POS_{B-\{a\}}(D) \neq POS_B(D)$.

Definition 3 implies two issues. One is that the reduct is jointly sufficient, the other is that the reduct is individually necessary for preserving a particular property (positive region in this context) of the decision systems [17]. The set of all relative reducts of S is denoted by $Red(S)$. The core of S is the intersection of these reducts, namely, $core(S) = \cap Red(S)$. Core attributes are of great importance to the decision system and should never be removed, except when information loss is allowed [37].

Throughout this paper, due to the positive region constraint, it is not necessary to construct a reduct. On the other side, we never want to select any redundant test. Therefore we propose the following concept.

Definition 4. *Let $S = (U, C, D, V, I)$ be a decision system. Any $B \subseteq C$ is a positive region sub-reduct of S iff $\forall a \in B$, $POS_{B-\{a\}}(D) \neq POS_B(D)$.*

According to the Definition 4, we observe the following:

(1) A reduct is also a sub-reduct.
(2) A core attribute may not be included in a sub-reduct.

Here we are interested those feature subsets satisfying the positive region constraint, and at the same time, with minimal possible test cost. To formalize the situation, we adopt the style of [18] and define the feature selection with positive region constraint problem, where the optimization objective is to minimize the test cost under the constraint.

Problem 1. Feature selection with positive region constraint (FSPRC) problem.
Input: $S = (U, C, d, V, I, c)$, the positive region lower bound pl;
Output: $B \subseteq C$;
Constraint: $|POS_B(D)|/|POS_C(D)| \geq pl$;
Optimization objective: $\min c(B)$.

In fact, the FSPRC problem is more general than the minimal test cost reduct problem, which is defined in [17]. In case where $pl = 1$, it coincides with the later. The minimal test cost reduct problem is in turn more general than the classical reduct problem, which is NP-hard. Therefore the FSPRC problem is at least NP-hard, and heuristic algorithms are needed to deal with it. Note that the FSPRC is different with the variable precision rough set model. The variable precision rough set model changes the lower approximation by varying the accuracy, but in our problem definition, it is unchanged.

3   The Algorithm

Similar to the heuristic algorithm to the OSRT problem [20], we also design a heuristic algorithm to deal with the new problem. We firstly analyze the heuristic function which is the key issue in the algorithm. Let $B \subset C$ and $a_i \in C - B$, the information gain of a_i with respect to B is

$$f_e(B, a_i) = H(\{d\}|B) - H(\{d\}|B \cup \{a_i\}), \tag{3}$$

where $d \in D$ is a decision attribute. At the same time, the λ-weighted function is defined as

$$f(B, a_i, c, \lambda) = f_e(B, a_i)c_i^\lambda, \tag{4}$$

where λ is a non-positive number.

Algorithm 1. A heuristic algorithm to the FSPRC problem

Input: $S = (U, C, D, V, I, c)$, p_{con}, λ
Output: A sub-reduct of S
Method: FSPRC

1: $B = \emptyset$; //the sub-reduct
2: $CA = C$; //the unprocessed attributes
3: **while** $(|POS_B(D)| < p_{con})$ **do**
4:    For any $a \in CA$ compute $f(B, a, c, \lambda)$
      //Addition
5:    Select a' with maximal $f(B, a', c, \lambda)$;
6:    $B = B \cup \{a'\}$;
7:    $CA = CA - \{a'\}$;
8: **end while**
   //Deletion, B must be a sub-reduct
9: $CD = B$; //sort attribute in CD according to respective test cost in a descending order
10: **while** $CD \neq \emptyset$ **do**
11:    $CD = CD - \{a'\}$; //where a' is the first element in CD
12:    **if** $(POS_{B-\{a'\}}(D) = POS_B(D))$ **then**
13:       $B = B - \{a'\}$;
14:    **end if**
15: **end while**
16: return B

Our algorithm is listed in Algorithm 1. It contains two main steps. The first step contains lines 3 through 8. Attributes are added to B one by one according to the heuristic function indicated in Eq. (4). This step stops while the positive region reaches the lower bound. The second step contains lines 9 through 15. Redundant attributes are removed from B one by one until all redundant have been removed. As discussed in Sect. 2.3, our algorithm has not a stage of core computing.

4   Experiments

To study the effectiveness of the algorithm, we have undertaken experiments using our open source software Coser [24] on 4 different datasets, i.e., Zoo, Iris,

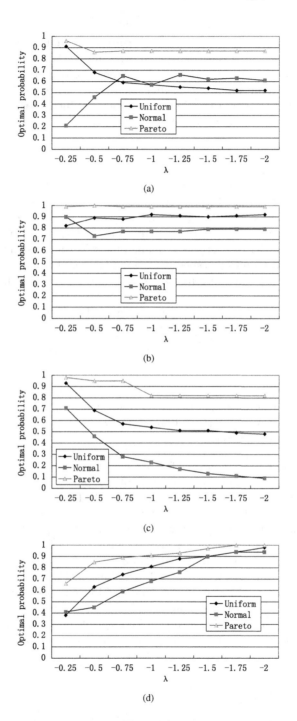

Fig. 1. Optimal probability: (a) zoo; (b) iris; (c) voting; (d) tic-tac-toc.

Voting, and Tic-tac-toe, downloaded from the UCI library [5]. To evaluate the performance of the algorithm, we need to study the quality of each sub-reduct which it computes. This experiment should be undertaken by comparing each sub-reduct to an optimal sub-reduct with the positive region constraint. Unfortunately, the computation of an optimal sub-reduct with test positive region constraint is more complex than that of a minimal reduct, or that of a minimal test cost reduct. In this paper, we only study the influence of λ to the quality of the result.

4.1 Experiments Settings

Because of lacking the predefined test costs in the four artificial datasets, we specify them as the same setting as that of [17] to produce test costs within [1, 100]. Three distributions, namely, Uniform, Normal, and bounded Pareto, are employed. In order to control the shape of the Normal distribution and the bounded Pareto distribution respectively, we must set the parameter α. In our experiment, for the Normal distribution, $\alpha = 8$, and test costs as high as 70 and as low as 30 are often generated. For the bounded Pareto distribution, $\alpha = 2$, and test costs higher than 50 are often generated. In addition, we intentionally set the constraint as $pl = 0.8$. This setting shows that we need a sub-reduct rather than a reduct.

4.2 Experiments Results

The experimental results of the 4 datasets are illustrated in Fig 1. By running our program in different λ values, the 3 different test cost distributions are compared. We can observe the following.

(1) The algorithm finds the optimal feature subset in most cases. With appropriate settings, it achieves more than 70 % optimal probability on these datasets.
(2) The result is influenced by the user-specified λ. The probability of obtained the best results is different with different λ values, where the "best" means the best one over the solutions we obtained, not the optimal one.
(3) The algorithm's performance is related with the test cost distribution. It is best on datasets with bounded Pareto distribution. At the same time, it is worst on datasets with Normal distribution. Consequently, if the real data has test cost subject to the Normal distribution, one may develop other heuristic algorithms to this problem.
(4) There is not a setting of λ such that the algorithm always obtain the best result. Therefore the settings might be learned instead of provided by the user.

5 Conclusions

In this paper, we firstly proposed the FSPRC problem. Then we designed a heuristic algorithm to deal with it. Experimental results indicate that the optimal solution is not easy to obtain. In the future, one can borrow ideas from [13, 30, 33, 39] to develop an exhaustive algorithm to evaluate the performance of a heuristic algorithm. One also can develop more advanced heuristic algorithms to obtain better performance.

Acknowledgements. This work is partially supported by the Natural Science Foundation of Department of Education of Sichuan Province under Grant No. 13ZA0136, and National Science Foundation of China under Grant Nos. 61379089, 61379049.

References

1. Chen, X.: An improved branch and bound algorithm for feature selection. Pattern Recogn. Lett. **24**(12), 1925–1933 (2003)
2. Dash, M., Liu, H.: Feature selection for classification. Intell. Data Anal. **1**, 131–156 (1997)
3. Fayyad, U., Piatetsky-Shapiro, G., Smyth, P.: From data mining to knowledge discovery in databases. AI Mag. **17**, 37–54 (1996)
4. Greco, S., Matarazzo, B., Slowinski, R., Stefanowski, J.: Variable consistency model of dominance-based rough sets approach. In: Ziarko, W.P., Yao, Y. (eds.) RSCTC 2000. LNCS (LNAI), vol. 2005, pp. 170–181. Springer, Heidelberg (2001)
5. Blake, C.L., Merz, C.J.: UCI repository of machine learning databases (1998). http://www.ics.uci.edu/~mlearn/mlrepository.html
6. He, H.P., Min, F.: Accumulated cost based test-cost-sensitive attribute reduction. In: Kuznetsov, S.O., Ślęzak, D., Hepting, D.H., Mirkin, B.G. (eds.) RSFDGrC 2011. LNCS, vol. 6743, pp. 244–247. Springer, Heidelberg (2011)
7. He, H.P., Min, F., Zhu, W.: Attribute reduction in test-cost-sensitive decision systems with common-test-costs. In: Proceedings of the 3rd International Conference on Machine Learning and Computing, vol. 1, pp. 432–436 (2011)
8. Hu, Q.H., Yu, D.R., Liu, J.F., Wu, C.: Neighborhood rough set based heterogeneous feature subset selection. Inf. Sci. **178**(18), 3577–3594 (2008)
9. Hunt, E.B., Marin, J., Stone, P.J. (eds.): Experiments in Induction. Academic Press, New York (1966)
10. Lanzi, P.: Fast feature selection with genetic algorithms: a filter approach. In: IEEE International Conference on Evolutionary Computation 1997. IEEE (1997)
11. Lin, T.Y.: Granular computing on binary relations - analysis of conflict and Chinese wall security policy. In: Alpigini, J.J., Peters, J.F., Skowron, A., Zhong, N. (eds.) RSCTC 2002. LNCS (LNAI), vol. 2475, pp. 296–299. Springer, Heidelberg (2002)
12. Lin, T.Y.: Granular computing - structures, representations, and applications. In: Wang, G., Liu, Q., Yao, Y., Skowron, A. (eds.) RSFDGrC 2003. LNCS (LNAI), vol. 2639, pp. 16–24. Springer, Heidelberg (2003)
13. Liu, Q.H., Li, F., Min, F., Ye, M., Yang, W.G.: An efficient reduction algorithm based on new conditional information entropy. Control Decis. (in Chinese) **20**(8), 878–882 (2005)
14. Liu, J.B., Min, F., Liao, S.J., Zhu, W.: A genetic algorithm to attribute reduction with test cost constraint. In: Proceedings of 6th International Conference on Computer Sciences and Convergence Information Technology, pp. 751–754 (2011)
15. Liu, H., Motoda, H.: Feature Selection for Knowledge Discovery and Data Mining. The Springer International Series in Engineering and Computer Science, vol. 454. Kluwer Academic Publishers, Boston (1998)

16. Ma, L.W.: On some types of neighborhood-related covering rough sets. Int. J. Approx. Reason. **53**(6), 901–911 (2012)
17. Min, F., He, H.P., Qian, Y.H., Zhu, W.: Test-cost-sensitive attribute reduction. Inf. Sci. **181**, 4928–4942 (2011)
18. Min, F., Hu, Q.H., Zhu, W.: Feature selection with test cost constraint. Int. J. Approximate Reasoning (2013, to appear). doi:10.1016/j.ijar.2013.04.003
19. Min, F., Liu, Q.H.: A hierarchical model for test-cost-sensitive decision systems. Inf. Sci. **179**, 2442–2452 (2009)
20. Min, F., Zhu, W.: Attribute reduction with test cost constraint. J. Electr. Sci. Technol. China **9**(2), 97–102 (2011)
21. Min, F., Zhu, W.: Minimal cost attribute reduction through backtracking. In: Kim, T., et al. (eds.) DTA/BSBT 2011. CCIS, vol. 258, pp. 100–107. Springer, Heidelberg (2011)
22. Min, F., Zhu, W.: Optimal sub-reducts in the dynamic environment. In: Proceedings of IEEE International Conference on Granular Computing, pp. 457–462 (2011)
23. Min, F., Zhu, W.: Optimal sub-reducts with test cost constraint. In: Yao, J.T., Ramanna, S., Wang, G., Suraj, Z. (eds.) RSKT 2011. LNCS, vol. 6954, pp. 57–62. Springer, Heidelberg (2011)
24. Min, F., Zhu, W., Zhao, H., Pan, G.Y., Liu, J.B., Xu, Z.L.: Coser: cost-sensitive rough sets (2012). http://grc.fjzs.edu.cn/~fmin/coser/
25. Pan, G.Y., Min, F., Zhu, W.: A genetic algorithm to the minimal test cost reduct problem. In: Proceedings of IEEE International Conference on Granular Computing. pp. 539–544 (2011)
26. Pawlak, Z.: Rough set approach to knowledge-based decision support. Eur. J. Oper. Res. **99**, 48–57 (1997)
27. Pawlak, Z.: Rough sets. Int. J. Comput. Inf. Sci. **11**, 341–356 (1982)
28. Pawlak, Z.: Rough sets and intelligent data analysis. Inf. Sci. **147**(12), 1–12 (2002)
29. Qian, Y.H., Liang, J.Y., Pedrycz, W., Dang, C.Y.: Positive approximation: an accelerator for attribute reduction in rough set theory. Artif. Intell. **174**(9–10), 597–618 (2010)
30. Skowron, A., Rauszer, C.: The discernibility matrices and functions in information systems. In: Intelligent Decision Support (1992)
31. Swiniarski, R.W., Skowron, A.: Rough set methods in feature selection and recognition. Pattern Recogn. Lett. **24**(6), 833–849 (2003)
32. Tseng, T.L.B., Huang, C.-C.: Rough set-based approach to feature selection in customer relationship management. Omega **35**(4), 365–383 (2007)
33. Wang, G.Y.: Attribute core of decision table. In: Alpigini, J.J., Peters, J.F., Skowron, A., Zhong, N. (eds.) RSCTC 2002. LNCS (LNAI), vol. 2475, pp. 213–217. Springer, Heidelberg (2002)
34. Wang, X., Yang, J., Teng, X., Xia, W., Jensen, R.: Feature selection based on rough sets and particle swarm optimization. Pattern Recogn. Lett. **28**(4), 459–471 (2007)
35. Xu, Z.L., Min, F., Liu, J.B., Zhu, W.: Ant colony optimization to minimal test cost reduction. In: Proceedings of the 2011 IEEE International Conference on Granular Computing. pp. 688–693 (2012)
36. Yao, J.T., Zhang, M.: Feature selection with adjustable criteria. In: Ślęzak, D., Wang, G., Szczuka, M.S., Düntsch, I., Yao, Y. (eds.) RSFDGrC 2005. LNCS (LNAI), vol. 3641, pp. 204–213. Springer, Heidelberg (2005)
37. Yao, Y.Y., Zhao, Y.: Attribute reduction in decision-theoretic rough set models. Inf. Sci. **178**(17), 3356–3373 (2008)

38. Zhang, W.X., Mi, J., Wu, W.: Knowledge reductions in inconsistent information systems. Chin. J. Comput. **26**(1), 12–18 (2003)
39. Zhao, H., Min, F., Zhu, W.: A backtracking approach to minimal cost feature selection of numerical data. J. Inf. Comput. Sci. **10**(13), 4105–4115 (2013)
40. Zhao, H., Min, F., Zhu, W.: Test-cost-sensitive attribute reduction based on neighborhood rough set. In: Proceedings of the 2011 IEEE International Conference on Granular Computing, pp. 802–806 (2011)
41. Zhao, H., Min, F., Zhu, W.: Test-cost-sensitive attribute reduction of data with normal distribution measurement errors. Math. Prob. Eng. **2013**, 12 pp (2013)
42. Zhong, N., Dong, Z.J., Ohsuga, S.: Using rough sets with heuristics to feature selection. J. Intell. Inf. Syst. **16**(3), 199–214 (2001)
43. Zhu, W.: Generalized rough sets based on relations. Inf. Sci. **177**(22), 4997–5011 (2007)
44. Zhu, W.: Topological approaches to covering rough sets. Inf. Sci. **177**(6), 1499–1508 (2007)
45. Zhu, W.: Relationship between generalized rough sets based on binary relation and covering. Inf. Sci. **179**(3), 210–225 (2009)
46. Zhu, W., Wang, F.: Reduction and axiomization of covering generalized rough sets. Inf. Sci. **152**(1), 217–230 (2003)

A Rough Neurocomputing Approach for Illumination Invariant Face Recognition System

Singh Kavita[1](✉), Zaveri Mukesh[2], and Raghuwanshi Mukesh[3]

[1] Computer Technology Department, Y.C.C.E, Nagpur 441110, India
singhkavita19@yahoo.co.in
[2] Computer Engineering Department, S.V.N.I.T, Surat 329507, India
mzaveri@coed.svnit.ac.in
[3] Rajiv Gandhi College of Engineering and Research, Nagpur 441110, India
m_raghwanshi@rediffmail.com

Abstract. To surmount the issue of illumination variation in face recognition, this paper proposes a rough neurocomputing recognition system, namely, RNRS for an illumination invariant face recognition. The main focus of the proposed work is to address the problem of variations in illumination through the strength of the rough sets to recognize the faces under varying effects of illumination. RNRS uses geometric facial features and an approximation-decider neuron network as a recognizer. The novelty of the proposed RNRS is that the correct face match is estimated at the approximation layer itself based on the highest rough membership function value. On the contrary, if it is not being done at this layer then decider neuron does this against reduced number of sample faces. The efficiency and robustness of the proposed RNRS are demonstrated on different standard face databases and are compared with state-of-art techniques. Our proposed RNRS has achieved 93.56 % recognition rate for extended YaleB face database and 85 % recognition rate for CMU-PIE face database for larger degree of variations in illumination.

Keywords: Rough sets · Neurocomputing · Illumination variation · Face recognition

1 Introduction

The presence of illumination variation changes the appearance of the face and further effects the performance of face recognition. To address this issue, several illumination invariant face recognition systems with different illumination normalization techniques and classifiers have been proposed [1,2,4–9]. These classifiers can be broadly categorised into distance metric-based [8], probabilistic-based [4–7] and learning-based [9–11] algorithms. Most widely distance metric-based classifier used is nearest neighbor classifier (NNC) [8] because of its simplicity. But the disadvantage is that the performance of NNC degrades

© Springer-Verlag Berlin Heidelberg 2014
J.F. Peters et al. (Eds.): Transactions on Rough Sets XVIII, LNCS 8449, pp. 34–51, 2014.
DOI: 10.1007/978-3-662-44680-5_3

for larger degree of illumination variation. Secondly, the other category that is probabilistic-based classifiers which includes Bayesian classifier [3,4] and probabilistic networks [5–7]. Third category, *i.e.,* learning-based classifiers such as classical neural network [10,11] and radial basis function network (RBF) [9] have been used widely for face recognition. A neural network has the ability to learn different patterns for recognition by adjusting weights from training data. The limitations of neural network-based approaches are that they require to define the network topology again when new training samples are added. It also increases the computational complexity.

Therefore, there is a need to develop a face recognition algorithm which will be efficient in the terms of learning ability, adaptability and computation. This shows that we need to incorporate the best characteristics of distance-based, probabilistic-based and learning-based algorithms discussed above. In this context, we propose a novel algorithm called as rough neurocomputing recognition system (RNRS). The novelty of RNRS is that it uses rough set theory (RST). To the best of our knowledge, RST has not been exploited for the complex problem of face recognition under illumination variation.

Proposed RNRS uses geometric facial features and an approximation-decider neuron network, namely, ADNN as a recognizer. The novelty of the proposed RNRS is that the correct face match is estimated at the approximation layer itself based on the highest rough membership function value. On the contrary, if it is not being done at this layer, then decider neuron does this against reduced number of sample faces. The efficiency and robustness of the proposed RNRS are demonstrated on different standard face databases and are compared with state-of-art techniques.

The paper is organized as follows: Sect. 2 briefly presents an overview of the rough set theory. Section 3 describes the different modules of proposed RNRS system that has been used for illumination invariant face recognition. Section 4 describes the simulation results. Section 5 concludes the paper.

2 Overview of Rough Set Theory

In literature, uncertainty has been defined as a measure of the users understanding of difference between the contents of a dataset representing certain phenomena. For example, in our case, uncertainty in face recognition due to illumination variations arises since the varying effects of illumination on facial images change the appearances of faces of the same subject captured under different illuminations. In this view, in general it can be said that from the acquired data in a dataset, it is not possible to discern the two different samples of the same object. RST takes into consideration the indiscernibility between objects typically characterized by an equivalence relation. Thus, rough sets are the results of approximating crisp sets using equivalence classes.

The different concepts of RST used in our work are presented as follows. RST is an approach put forth by Z. Pawlak [12]. It is used as a mathematical tool to handle the vagueness and uncertainty present in the data. In RST, information about the real world is given in the form of an information table. An information table can be represented as a pair $\mathcal{A} = (U, A)$, where, U is a non-empty finite set of objects called the universe and A is a non-empty finite set of attributes such that information function $f_a : U \rightarrow V_a$, for every $a \in A$. The set V_a is called the value set of a. Furthermore, a decision system is any information system of the form $\mathcal{A} = (U, A \cup d)$, where $d \notin A$ is a decision attribute. The main concept involves in RST is an indiscernibility relation. For every set of attributes $B \subseteq A$, an indiscernibility relation $IND(B)$ is defined in the following way: two objects, x_i and x_j, are indiscernible by the set of attributes $B \subseteq A$, if $b(x_i) = b(x_j)$ for every $b \subseteq B$. The equivalence class of $IND(B)$ is called elementary set in B because it represents the smallest discernible groups of objects. For any element x_i of U, the equivalence class of x_i in relation $IND(B)$ is represented as $[x_i]_{IND(B)}$. The notation $[x]_B$ denotes equivalence classes of $IND_A(B)$. The partitions induced by an equivalence relation can be used to build new subsets of the universe. The construction of equivalence classes is the first step in classification with rough sets.

Rough sets can also be defined by rough membership function instead of approximation. A rough membership function *(rmf)* makes it possible to measure the degree that any specified object with given attribute values belongs to a given decision set X. Let $B \subseteq A$ and x be a set of observations of interest. The degree of overlap between X and $[x]_B$ containing x can be quantified with a *rmf* given by Eq. (1).

$$\mu_x^B = \frac{|[x]_B \cap X|}{|[x]_B|} \tag{1}$$

$$\mu_x^B :\rightarrow [0,1] \tag{2}$$

where, $| \bullet |$ denotes the cardinality of a set. The rough membership value μ_x^B may be interpreted as the conditional probability that an arbitrary element x belongs to X given B. The decision set X is called a generating set of the rough membership μ_x^B. Thus rough membership function quantifies the degree of relative overlap between the decision set X and the equivalence class to which x belongs. In this view, the rough membership function can be used to define the boundary region of a set.

3   Proposed RNR System

For illumination invariant face recognition, we propose a novel approach based on RST called as rough neurocomputing-based recognition system (RNRS). RNRS consists of mainly three major modules which are also very common to any face recognition system. These modules consist of (1) pre-processing (2) feature extraction and (3) recognizer. Out of these three modules our major contribution is for pre-processing and recognition module. The novelty is that we exploit RST for both of these modules. We also use approximation decider neuron network, namely, ADNN [15] as recognizer which is very simple, effective and optimal in terms of performance and architectural point of view. The complete RNR system is shown in Fig. 1. Each module is described in detail as below.

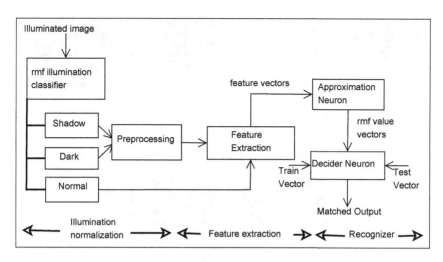

Fig. 1. Proposed RNRS System

3.1 Pre-processing

First module is pre-processing where various operations for enhancing the input image may be performed for better feature extraction. In our case, we perform illumination normalization as part of pre-processing. It is important to note that the illumination normalization step is applied in an adaptive manner. That is, the input image is first classified based on its illumination type or also called as image type into three categories, namely, *dark, shadow* and *normal.* For this illumination classification we deploy our novel *rmf*-based classifier [16,20]. If *rmf*-based classifier classify an input image either into *dark* or *shadow*, then illumination normalization technique is applied. In [16,20] the selection of illumination normalization technique is also discussed in detail which is very important as all the sample images of the same subject may have different degree of variation in illumination and if the same illumination normalization technique is used then it is not effective. It is also concluded in [16,20], that for *ambient* or *dark* images homomorphic technique [13,19] is suitable and self quotient image (SQI) [13,19] is more appropriate technique in case *shadow* images. For *normal* image there is no need to apply any illumination normalization technique which reduces the overall complexity of face recognition. This is the advantage of our proposed pre-processing step. Few of the classified and pre-processed sample images are from extended YaleB and CMU-PIE face databases are shown in Figs. 2 and 3 respectively.

3.2 Feature Extraction

The second module consists of feature extraction step. After illumination normalization, the input image is enhanced enough for extracting the features which are also crucial for the success of any face recognition. From the literature,

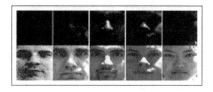

(a)*Shadow* images   (b)*Dark* images

Fig. 2. Classified and pre-processed sample images from extended YaleB database.

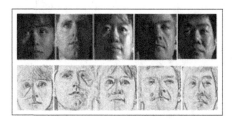

Fig. 3. Classified and pre-processed sample *shadow* images from CMU-PIE database

it is observed that large variety of features have been used for face recognition. Our main aim is to have good quality features which may represent face characteristic in a better way and should help recognizer. After studying the various feature sets, it is observed that geometrical features are more appropriate. The advantage of geometrical features extraction is that it helps in localizing different regions of a face like nose, ear, lips, and eyes which are very important for face representation. Other than this, the most advantage of selecting geometrical features is that it can be easily extracted after applying our adaptive illumination normalization in pre-processing step.

Various geometrical features that are used in our work are fiducial points such nose tip, eye corners and mouth corners. These fiducial points are enough to localize the face region. So it should be possible to extract these fiducial points from all the sample images of the same subject which might have varying degree of illumination. One observation from earlier discussion in Sect. 3.1 about different illumination normalization techniques is that selected illumination normalization methods successfully highlights fiducial points in all types of images irrespective of either it is *dark* or *shadow* or *normal*. These fiducial points may be detected manually or automatically from normalized images. In our work, we are interested to present our contribution for *rmf*-based classifier and recognition steps. So, currently, in our algorithm these fiducial points are marked manually presented in [14]. But it can be pursued by an automatic step without compromising the performance of our face recognition system. From the manually located, fiducial points, the distance between various pairs of fiducial points of eye, nose, lips *etc.,* are calculated and different angles formed by these fiducial points are also measured using simple distance formula.

The problem with these set of fiducial points is that it forms the large number of distance value pairs and the large number of possible angles which will increase into computational cost. To reduce this computational cost, we exploit the symmetry of the face depicted in Fig. 4. That is, the symmetry with respect to nose tip is considered by a perpendicular line if considered as mirror which divides the complete frontal face into left part and right part which are replica of each other. So, half number of distance value pairs and angles are redundant. This assumption further helps us when dealing with different poses later in our work. It means that only one part of face that is left or right should be sufficient for feature extraction for our proposed work.

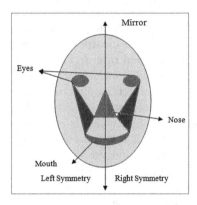

Fig. 4. Manually located fiducial points

Using these distance values and angles evaluated, a feature vector is formed. The distance between the fiducial points of eye corners and nose is named as Eye_nose_dist. Similarly the distance between nose tip and lip corners is represented by Nose_lip_dist. The list of such distance variables are mentioned with their symbolic representation in Table 1. Angle_EL represents the angle between the line joining (eye corner and the lip corner) and horizontal line. Angle_EN is the angle between the line joining the (eye corner and nose tip) and horizontal line. Angle_NL is the angle between the line joining the (nose tip and lip corner) and horizontal line.

These feature values form the attribute values set $B = \{b_i | 1 \cdots m\}$ where m is the total number of feature values. In our case B is defined as

$B=\{$Eye_nose_dist, Nose_lip_dist, Angle_EL, Angle_EN, Angle_NL$\}$ represented as

$B=\{b_1,b_2,b_3,b_4,b_5\}$, with $m=5$.

Once the feature vector is extracted from the localized face region for each input image, these feature vectors are used as an input to the next module *i.e.,* recognition step.

Table 1. Features description

Features name	Symbolic representation
Eye_nose_dist	b_1
Nose_lip_dist	b_2
Angle_EL	b_3
Angle_EN	b_4
Angle_NL	b_5

3.3   ADNN-Based Recognition

As mentioned at the beginning of this paper, the recognizer should be simple in terms of architecture, adaptive to incorporate the new training samples, and robust for achieving high recognition rate. In this regard, we use ADNN for face recognition in the presence of illumination variation. Earlier, ADNN has been used for fault detection in [15] which inspires us to exploit the simplicity and at the same time to have characteristics of three different types of recognizers for having a face recognition system which is efficient, robust and fast enough for real time application. ADNN is a two layer network in which the first layer consists of approximation neurons and the second layer consists of decider neuron. The approximation layer allows us to incorporate features of probabilistic and learning-based recognizers whereas the decider neuron uses the feature of distance-based recognizer. The complete design of ADNN is discussed in detail in the forthcoming section. The overview of complete ADNN-based face recognition is shown in Fig. 5.

3.4   Approximation Neuron

First stage of ADNN is approximation neuron layer. It consists of set of neurons called as approximation neurons. The number of approximation neurons is equal to the number of target class. In our proposed algorithm, the number of subjects are nothing but are the number of target classes. That is each target class is represented by one approximation neuron. This approximation neuron implements the rough membership function. The idea behind the mapping of rmf using an approximation neuron is that the rmf is nothing but the probabilistic measure of $x \in X$ given $IND[x]_B$. Here, x is a given object whose degree of membership is to be determined for a given set X. X is the set of objects. In our face recognition problem x is an input test image. X represent the set of images of a particular subject. So for each subject such X can be defined and we are interested to find out the degree of membership of a given test image x to each such X set. In our work, the degree of overlapping is measured using rmf.

The input to each approximation neuron is the set of feature vector $b_i, i = 1 \ldots d, b \subseteq B$ where B being the total number of attributes (features) under consideration described in Sect. 3.2. In this context we first define the set of objects U,

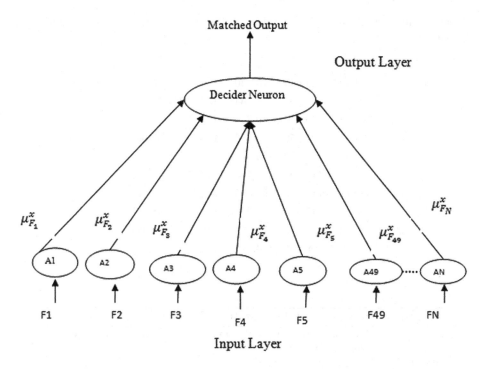

Fig. 5. ADNN architecture

and finite set of attributes $B \subseteq A$ to express the information table $\mathcal{A} = (U, A)$. For example, Table 2 shows an information table for 10 samples illuminated face images from extended YaleB face database. We assume each image $Y_i (i = 1, \cdots N)$ represent the object from the universe, where N is the number of sample facial images.

Next, we want to partition the given U into equivalence classes for rmf calculation. From Table 2 it is observed that attribute values are continuous values. These set of continuous values are proven to be rather unsuitable for depiction of crisp decision class and the attribute values have to be disjointedly discretized. Let us consider S be a set of continuous attribute value and let the domain of S be the interval $[p, q]$. A partition φ_S on $[p, q]$ will be partition defined as the following set of m subintervals shown in Eq. (3)

$$\varphi_S = \{[p_0, p_1), [p_1, p_2) \cdots [p_{m-1}, p_m]\} \tag{3}$$

where, $p_0 = p, p_{i-1} > p_i$, for $i = 1, 2 \cdots m$, and $p_m = q$. Thus, discretization is the process that produces a partition φ_S on $[p, q]$. Here, in our case, the discretization method used is depicted through the Algorithm 1.

The discretized values of attribute set B is redefined as function values $F = \{f_1, f_2, f_3, f_4, f_5\}$. With these discretized values represented as function values of objects in U, the discretized information table is created. A partial table is shown in Tables 3 and 4 for extended YaleB and CMU-PIE face databases,

Table 2. Information table with sample data for two subjects from extended YaleB database

Objects	b_1	b_2	b_3	b_4	b_5
Y_1	76.059	104.695	73.430	135.473	46.581
Y_2	78.230	104.403	68.264	139.170	48.527
Y_3	79.649	105.342	69.836	138.479	49.117
Y_4	82.855	107.856	73.783	136.918	50.088
Y_5	84.723	105.688	66.573	140.992	53.228
Y_6	83.600	115.694	72.719	141.250	46.021
Y_7	89.811	116.760	68.505	144.180	50.107
Y_8	87.006	115.802	68.264	144.086	48.385
Y_9	80.784	103.719	73.573	148.496	46.421
Y_{10}	88.238	109.895	64.815	143.859	53.408
$\vdots$	$\vdots$	$\vdots$	$\vdots$	$\vdots$	$\vdots$

Algorithm 1. Interval-based discretization of continuous attribute values

Input: $N \times P$ continuous attribute value matrix, N = number of samples,
$\qquad$ P = number of attributes
Output: Discretized information table

```
1  for i ← 1 to N do
2      for j ← 1 to P do
3          find the interval range for Pⱼ;
4          label each Pⱼ with l = 1, 2, 3, ... n;
5      create information table for Nᵢ
6  return discretized information table;
```

---

respectively. In addition, the facial images present in training set can be used to decide the values for decision attribute d in creating the decision table. It enables in deciding the decision or target class for the test images.

With this framework of an approximation layer, each approximation neuron evaluates rmf value for each target class. The rmf value ranges from 0 to 1. Indirectly, the rmf value provides matching measure of a given test input image belongs to a particular class *i.e.*, the subject for recognition. By mapping rmf through neuron we are benefited by two major points: (i) rmf is based on RST and as discussed in earlier chapter, the rough set is preferred to handle imprecision and vagueness in the information and (ii) rmf is a probabilistic measure as pointed out and hence, the given problem is solved using statistical approach.

Table 3. Extended YaleB face database   **Table 4.** CMU-PIE face database

Objects	f_1	f_2	f_3	f_4	f_5	d
Y_1	1	5	5	1	1	1
Y_2	1	5	4	1	1	1
Y_3	1	6	4	1	1	1
Y_4	2	6	5	1	1	1
Y_5	2	6	4	2	2	1
Y_6	2	7	5	2	1	2
Y_7	3	7	4	2	1	2
Y_8	2	7	4	2	1	2
Y_9	1	5	5	3	1	2
Y_{10}	3	6	4	2	2	2
$\vdots$	$\vdots$	$\vdots$	$\vdots$	$\vdots$	$\vdots$	$\vdots$

Objects	f_1	f_2	f_3	f_4	f_5	d
C_1	4	5	4	3	2	1
C_2	3	5	5	2	2	1
C_3	3	4	4	2	2	1
C_4	3	4	4	3	3	1
C_5	4	5	4	3	3	1
C_6	3	1	1	4	0	2
C_7	3	1	1	3	0	2
C_8	3	1	1	3	0	2
C_9	4	2	1	4	0	2
C_{10}	2	1	1	3	0	2
$\vdots$	$\vdots$	$\vdots$	$\vdots$	$\vdots$	$\vdots$	$\vdots$

3.5   Decider Neuron

The approximation layer neurons provide the degree of overlapping that is measure of matching in terms of rmf values between input test image and different

---

Algorithm 2. Decider Neuron

Input: Feature matrix of face samples having equal $rmfs$, and test feature
      values of P
Output: Index of matched identity
1   **for** $i \leftarrow 1$ *to* N **do**
2   **if** $max\ \{\ rmfN_i\ \}$ **then**
3
4   assign P to subject class C_i
5   **return** *Index i of matched face*
6   **else**
7
8   count the occurrences of $\{rmfs\ \}$ having equal values
9   **if** *count ¿ 1* **then**
10   find reduced feature matrix for number of face samples n_j having equal
       rmfs, $n \subset N$
11   $bestmatch = min$
12   **for** $j \leftarrow 1$ *to* n **do**
13   compute $diff_j$ using Euclidean distance $w.r.t$ test face
14   **if** $diff_j$ ¿ *best match* **then**
15   assign P to subject class C_j
16   **return** *Index j of matched face*
17 **End**

target classes. Higher value of rmf depicts the matching. Now, to assign an input test image to a particular target class for final face recognition, the rmf values obtained from the approximation layers are compared with a predetermined threshold value. This threshold value may be selected based on the application. The set of target classes for which rmf values are higher than the threshold are further sent to decider neuron for finding an exact target class which can be assigned to test image. For each of target class in this set, the feature vector derived from input test image is compared with all feature vectors of respective class. This comparison is performed using standard distance-based technique which is implemented by decider neuron. For simplicity and fast execution, Euclidean distance measure is used for comparing two feature vectors. The input test image is assigned to a particular target class whose feature vector is providing minimum distance with that of input test image. Further, we also describe the designing of decider neuron in Algorithm 2.

As discussed earlier, the input to decider neuron is the matrix of rmf values computed for each test facial image with respect to each approximation neuron. From all the rmf values given to decider neuron, it assigns the test facial image P to a specific subject face class C_i, $i = 1 \ldots K$ with the highest rmf values, where, $K \subseteq N$, where N is the total number of subject/target classes. If the test facial image has been approximated to more than one target classes due to $max\{rmf_k\} = rmf_i = rmf_j = rmf_k$, $i, j, k \subseteq K$, provided $i \neq j \neq k$ can be resolved by decider neuron.

In this way the distance-based recognizer is exploited in our approach at second stage. It is important to note that our computational complexity is reduced at both the stages of ADNN framework. First level reduces the search space using equivalence classes and at second level only subset of target classes is exploited for distance-based comparison. Overall discussion shows that the proposed RNRS using ADNN is much more effective in the terms of architecture, adaptively, simplicity and efficient in the terms of computations compared to other categories of recognizers which is also evident from the simulations results described next.

4   Simulation Results

For simulation, extended YaleB [17] and CMU-PIE [18] face databases are used. Extended YaleB consists of whole range of illumination space and has obviously become a standard for variable-illumination recognition algorithms. The extended YaleB face database contains approximately 2496 facial images (640 × 486 pixels) of 39 different individual subjects. On the other hand, CMU-PIE face database consists of 41,368 images of 68 subjects. The images are acquired from 13 different poses, with 43 different illumination conditions and with 4 different facial expressions. The images are (640 × 486 pixels) colour images. Since we mainly focus on illumination problem, only frontal images are considered that are captured from camera c27. Moreover, this database also consists of images captured both with the room lights on as shown in Fig. 6(a) and with the room lights off as shown in Fig. 6(b).

Fig. 6. Samples facial images from CMU-PIE face database

On the contrary extended YaleB face database consists of the images that are captured with the room lights switched off. Therefore, only those facial images are selected from CMU-PIE face database which are similar to the extended YaleB face database in order to validate the efficacy of RNRS to illumination change for both the facial databases. To remove the irrelevant information faces are manually cropped from the images and resized to 128 × 128 pixels. Sample cropped and resized facial images from both the face databases, *i.e.*, YaleB and CMU-PIE are depicted in Fig. 7.

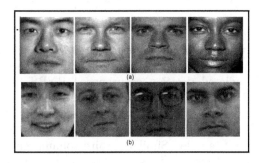

Fig. 7. Sample cropped and resized *normal* images to 128 × 128 pixels; (a) extended YaleB (b) CMU-PIE

Following cross-validation method, face databases are partitioned into training set and testing set. For each face database, training set consists of randomly selected 100 images for 20 different individuals with five face samples per individual. RNRS has been trained with only *normal* images that have been classified by *rmf*-based classifier used for illumination classification. Recognition performance of the proposed RNRS have been evaluated on 120 facial images of 20 individuals with varying illumination range in the azimuth and elevation angle with 6 face samples per individual. Illuminated images are first classified forming three separated test subsets: set1, set2 and set3. Set1 is the set of classified *normal* images; set2 consists of classified *shadow* images and set3 is of classified *dark* images.

At this point, it is to be noted that, selection of training images greatly affects the performance of face recognition system. Therefore, we performed the training with training sets consisting different number of samples per subject with $S=1$, $S=3$ and $S=5$. (S is number of training sample per subject). The respective recognition rate achieved for each training set is depicted in Table 5. From Table 5, it is clear that recognition rate improves to a great extent if number of samples per subject increases. This is due to the fact that more number of feature values can be extracted from more number of samples that provide more accurate training model. Therefore, we continued the training of the proposed recognition system with $S=5$ because even if the number of training samples per subject was increased more than five, no improvement in recognition rate was observed.

Table 5. Recognition rates in percentage for varying number of samples per subject

Illumination type	Recognition rates		
Normalization technique	$S=1$	$S=3$	$S=5$
NORMAL	50.75	80.23	98.18
SHADOW+SQI	45.00	76.02	92.51
DARK+HOMOMORPHIC	43.00	75.10	90.00

Snapshots of the resultant output of proposed RNRS are depicted in Tables 6 and 7 for extended YaleB and CMU-PIE face databases respectively. Presented tables depict the predicted decision of proposed RNRS for test images for extended YaleB and CMU-PIE face databases. In Table 6, first column represents the object number or test image number (TIN) for test samples. It is represented as T_i ($i = 1 \cdots N$), N is the total number of test samples. Second column depicts the actual target face of test samples. Similarly, third column represents the *normal* test images. Each *normal* test image is shown with function values. And, lastly the fourth column is for predicted decision of proposed RNRS. First and second column is common for *shadow* and *dark* test images shown as separate column in the same table. In the same context Table 7 is also defined.

From both the tables, predicted value of proposed RNRS for each test face can be compared against the target face value from second column. This comparison can help in taking the decision whether the test image has been successful matched or not. For instance, from Table 6 it is clear that *dark* test images T_3 and T_4 from seventh column from left to right are of subject number 2. It is clear from the eighth column of Table 6, that although test image T_3 belongs to subject number 2 (target number 2) but it has been classified as 39 *i.e.,* the predicted face number is 39. This is an example of misclassification. On the other hand, T_4 has been correctly matched with target face value, *i.e.,* 2. Similarly, through the confusion matrix, overall recognition rates for each classifier are evaluated.

The performance of proposed RNRS is also compared with other approaches such as nearest neighbour and Naive Bayes classifiers. The aforementioned classifiers are implemented as the standard algorithms for comparison. Summarized face recognition performances on extended YaleB and CMU-PIE face databases are tabulated in Tables 8 and 9 respectively.

Table 6. Snapshot of resultant table of proposed RNRS for extended YaleB database (TIN: Test Image Number, TF: Target Face, PMPR: Predicted Match Class by Proposed RNRS)

TIN	TF	Normal	PMPR	Shadow	PMPR	Dark	PMPR
T_1	1	1 5 4 1 1	1	2 6 5 1 1	1	1 5 5 1 1	1
T_2	1	1 6 4 1 1	1	1 5 4 1 1	1	1 5 5 1 1	1
T_3	2	3 7 4 2 1	2	3 7 4 2 1	2	3 6 4 2 1	39
T_4	2	3 7 4 2 1	2	3 7 4 2 1	2	3 7 4 2 1	2
T_5	3	2 6 2 3 2	3	3 6 2 3 2	3	3 6 2 3 2	3
T_6	3	2 5 4 2 2	3	3 6 2 3 2	3	3 6 2 3 2	3
T_7	4	1 4 3 2 2	4	1 4 3 2 2	4	1 4 3 2 2	4
T_8	4	1 5 0 1 1	4	1 4 3 2 2	4	1 4 3 2 2	4
T_9	5	5 7 2 3 3	5	3 4 3 2 4	22	3 6 3 2 2	31
T_{10}	5	5 7 2 4 3	5	3 4 2 2 0	39	5 5 2 3 0	39
⋮	⋮	⋮	⋮	⋮	⋮	⋮	⋮

Table 7. Snapshot of resultant table for proposed RNRS using CMU-PIE face database (TIN: Test Image Number, TF: Target Face, PMPR: Predicted Match Class by Proposed RNRS)

TIN	TF	Normal	PMPR	TF	Shadow	PMPR
T_1	1	3 5 5 2 2	1	1	3 4 4 3 3	1
T_2	1	4 5 4 3 3	1	1	4 4 7 0 3	5
T_3	1	3 4 4 2 2	2	1	4 5 4 3 3	1
T_4	2	4 2 1 4 0	2	2	3 1 1 3 0	2
T_5	2	2 1 1 3 0	3	2	2 1 1 3 0	2
T_6	3	5 5 7 1 2	17	2	3 1 1 3 0	2
T_7	3	5 6 7 1 2	11	2	4 3 6 1 3	20
T_8	4	2 4 7 0 1	10	3	5 6 7 1 3	3
T_9	4	1 2 6 0 2	5	3	5 5 7 1 2	3
T_{10}	5	4 3 6 0 4	5	3	3 1 1 4 0	2
⋮	⋮	⋮	⋮	⋮	⋮	⋮

Table 8. Face recognition rates (%) for different classifiers using extended YaleB database

Illumination type	Nearest neighbor	Naive Baye's	Proposed RNRS
Normal	98.18	67.27	98.18
Shadow	60.00	57.05	92.51
Dark	52.15	55.00	90.00

From the numerical figures summarized in Table 8, it is evident that, although the performance of nearest neighbor is equivalent to proposed RNRS for *normal* images, performance degrades when *shadow* and *dark* images are compared against the *normal* images. On the contrary, proposed RNRS outperforms the other two classifiers even in cases of *shadow* and *dark* images.

Table 9. Face recognition rates (%) for different classifiers using CMU-PIE database

Illumination type	Nearest neighbour	Naive Baye's	Proposed RNRS
Normal	98	88	90
Shadow	68	64	80

From the results in Tables 8 and 9, the overall performance of proposed RNRS is computed as 93.56 % for extended YaleB and 85 % CMU-PIE face databases, respectively. These results very clearly depict that proposed RNRS has shown better results using extended YaleB face database that consists wide range of illumination variations. Performance using CMU-PIE is tabulated only for *normal* and *shadow* images as there is no *dark* images in CMU-PIE face database. The performance for CMU-PIE face database is inferior compared to extended YaleB since CMU-PIE consists of images of different sizes of same person and the template we used for geometrical feature extraction are of fixed size which results in less recognition rate. It is evident from the Table 9 that, the performance of the proposed RNRS is not much affected, even when *shadow* images are compared with the *normal* images.

Further, the number of true positives and false positives are also plotted for each classifier under comparison as shown in Fig. 8. Plots in Fig. 8(b) depict that only one false positive is there for proposed RNRS and nearest neighbor and so the curve exactly fit with the target face curve, where as for Naive Baye's classifier there are more numbers of false positives and so the output curve is not in line with the target face curve. It is clear from all the plots shown in Fig. 8, that proposed RNRS has outperformed among two classifiers for both the face databases and robust against all type of variations in illumination. The proposed RNRS is also evaluated on AT&T [21] face database. The AT&T face database consists of 400 images of 40 subjects. All the face images are of size 92×112 with 256 gray levels. The databases have male and female subjects, with or without glasses, with or without beard, with or without some facial expression with varying effect of illuminations. Training is performed with randomly chosen 5 samples per subject and remaining 5 images are used for testing. The plot shown in Fig. 8(a) depicts the true positives and false positives for AT&T face database and the overall recognition rate is evaluated as 99 % which is excellent in terms of recognition rate. But, we have only consider the performance of extended YaleB and CMU-PIE face databases for discussion as these two face databases consists of larger degree of variations in illumination.

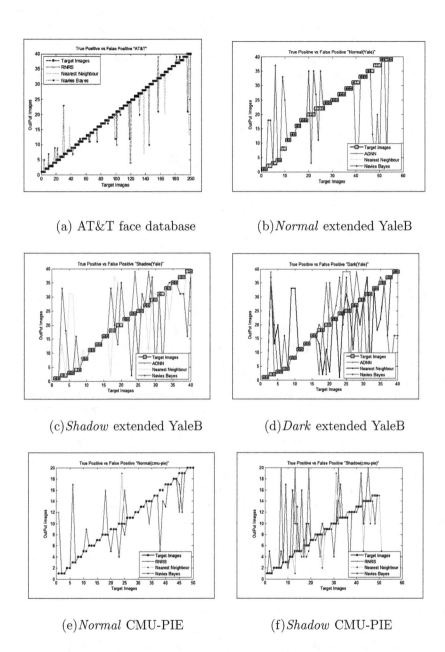

(a) AT&T face database   (b) *Normal* extended YaleB

(c) *Shadow* extended YaleB   (d) *Dark* extended YaleB

(e) *Normal* CMU-PIE   (f) *Shadow* CMU-PIE

Fig. 8. Plots showing the true positives and false positives.

We also compared the performance of proposed RNRS with standard PCA [2] and LDA [2] algorithms. From Table 10, it is clear that proposed RNRS system has no doubtly outperformed these two algorithms. The improvement in the performance of RNRS to be noted is approximately by 20~30 % for both CMU-PIE and extended YaleB face databases. Another important point that can be observed from the results tabulated in Table 10 that these two algorithms PCA and LDA are not too robust to variations in illumination.

Table 10. Overall comparative recognition rates for different approaches

Approches	Extended YaleB	CMU-PIE
PCA	57.2 %	64.56 %
LDA	59.38 %	74.25 %
Proposed RNRS	93.56 %	85 %

5   Conclusion

This paper has presented an illumination invariant face recognition system using RST called as RNRS. Proposed RNRS is capable in tackling the larger degree of variations in illumination and robust in recognizing the faces under such larger degree of variations. Robustness is achieved by using the concept of rough membership function from rough sets to cope with illumination variations that inhibit the performance RNRS. Proposed RNRS has been compared with state-of-art classifiers studied in the literature. On an average, the proposed RNRS system has shown 93.56 % recognition rate for extended YaleB face database and 85 % recognition rate for CMU-PIE face database.

There is a scope of several improvements to proposed RNRS system. The first and the important one is by employing some automatic feature extraction technique with RNRS. Another possible enhancement is to do face recognition using ADNN recognizer under variations in poses.

Acknowledgement. This project is supported by AICTE of India (Grant No: 8023/RID/RPS-81/2010-11, Dated: March 31, 2011).

References

1. Turk, M., Pentland, A.: Eigenfaces for recognition. J. Cogn. Neurosci. **3**, 71–86 (1991)
2. Belhumeur, P.N., Hespanha, J.P., Kriegman, D.J.: Eigenfaces vs. Fisherfaces: recognition using class specific linear projection. IEEE Trans. Pattern Anal. Mach. Intell. **19**, 711–720 (1997)

3. Xiaogang, W.S., Xiaoou, T.: Bayesian face recognition using gabor features. In: Proceedings of the ACM SIGMM Workshop, pp. 70–73 (2003)
4. Alpaydin, E.: Introduction to Machine Learning. MIT press, Cambridge (2004)
5. Yiu, K., Mak, M., Li, C.: Gaussian Mixture Models and Probabilistic Decision-Based Neural Networks for Pattern Classification: A Comparative Study. Hong Kong Polytechnic University, Hong Kong (1999)
6. Moghaddam, B., Pentland, A.: Probabilistic visual learning for object representation. IEEE Trans. Pattern Anal. Mach. Intell. **19**, 696–710 (1997)
7. Liu, C., Wechsler, H.: Probabilistic reasoning models for face recognition. In: Proceedings of Computer Vision and Pattern Recognition, pp. 827–832 (1998)
8. Sang-II, C., Chong-Ho, C.: An effective face recognition under illumination and pose variation. In: Proceedings of International Joint Conference on Neural Networks, pp. 914–919 (2007)
9. Howell, A., Buxton, H.: Face recognition using radial basis function neural networks. In: Proceedings of the British Machine Vision Conference, pp. 455–464 (1996)
10. Rowley, H., Baluja, S., Kanade, T.: Neural network-based face detection. IEEE Trans. Pattern Anal. Mach. Intell. **20**, 23–38 (1998)
11. Garica, C., Delakis, M.: Convolution face finder: a neural architecture of fast and robust face detection. IEEE Trans. Pattern Anal. Mach. Intell. **26**, 1408–1423 (2004)
12. Pawlak, Z.: Rough sets. Int. J. Comput. Inform. Sci. **11**(5), 341–356 (1982)
13. Wang, H., Li, S.Z., Wang, Y.: Face recognition under varying lightening conditions using self quotient image. In: Proceedings of IEEE International Conference on Automatic Face and Gesture Recognition (FGR), pp. 819–824 (2004)
14. Kavita, S.R., Mukeshl, Z.A., Mukesh, R.M.: Extraction of pose invariant facial features. In: Das, V.V., Vijaykumar, R. (eds.) ICT 2010. CCIS, vol. 101, pp. 535–539. Springer, Heidelberg (2010)
15. Han, L., Peters, J.F.: Rough neural fault classification of power system signals. In: Peters, J.F., Skowron, A. (eds.) Transactions on Rough Sets VIII. LNCS, vol. 5084, pp. 396–519. Springer, Heidelberg (2008)
16. Singh, K.R., Zaveri, M.A., Raghuwanshi, M.M.: A robust illumination classifier using rough sets. In: Proceedings of Second IEEE International Conference on Computer Science and Automation Engineering, pp. 10–12 (2011)
17. http://cvc.yale.edu/projects/yalefaces/yalefaces.html
18. Sim, T., Baker, S., Bsat, M.: The CMU pose, illumination, and expression database. IEEE Trans. Pattern Anal. Mach. Intell. **25**, 1615–1618 (2003)
19. de Solar, J.R., Quinteros, J.: Illumination compensation and normalization in eigen space-based face recognition: a comparative study of different pre-processing approaches. Comput. Vis. Graph. Image Process. **42**, 342–350 (1999)
20. Singh, K.R., Zaveri, M.A., Raghuwanshi, M.M.: Rough membership function-based illumination classifier for illumination invariant face recognition. Elsevier Appl. Soft Comput. **13**(10), 4105–4117 (2013)
21. AT & T or ORL Face Database (2010). http://www.uk.research.att.com/facedatabase.html. Accessed July 2010

Variable Precision Multigranulation Rough Set and Attributes Reduction

Hengrong Ju[1,2], Xibei Yang[1,3,4(✉)], Huili Dou[1,2], and Jingjing Song[1,2]

[1] School of Computer Science and Engineering,
Jiangsu University of Science and Technology,
Zhenjiang 212003, Jiangsu, People's Republic of China
[2] Key Laboratory of Intelligent Perception and Systems for High-Dimensional
Information, Nanjing University of Science and Technology, Ministry of Education,
Nanjing 210094, People's Republic of China
[3] Artificial Intelligence Key Laboratory of Sichuan Province,
Zigong 643000, People's Republic of China
[4] School of Economics and Management,
Nanjing University of Science and Technology,
Nanjing 210094, People's Republic of China
yangxibei@hotmail.com

Abstract. Multigranulation rough set is a new expansion of the classical rough set since the former uses a family of the binary relations instead of single one for the constructing of approximations. In this paper, the model of the variable precision rough set is introduced into the multigranulation environment and then the concept of the variable precision multigranulation rough set is proposed, which include optimistic and pessimistic cases. Not only basic properties of variable precision multigranulation rough set are investigated, but also the relationships among variable precision rough set, multigranulation rough set and variable precision multigranulation rough set are examined. Finally, a heuristic algorithm is presented for computing reducts of variable precision multigranulation rough set, it is also tested on five UCI data sets.

Keywords: Attributes reduction · Multigranulation rough set · Variable precision multigranulation rough set

1 Introduction

Rough set [1], proposed by Pawlak, is a powerful tool, which can be used to deal with the inconsistency problems by separation of certain and uncertain decision rules from the exemplary decisions.

In Pawlak's rough set model, the classification criterion should be correct. However, in many practical applications, it seems that admitting some levels of errors in the classification processing may lead to a deeper understanding and a better utilization of properties of the data being analyzed. To achieve such goal,

---

This is an extended version of the paper presented at the 2012 Joint Rough Set Symposium.

© Springer-Verlag Berlin Heidelberg 2014
J.F. Peters et al. (Eds.): Transactions on Rough Sets XVIII, LNCS 8449, pp. 52–68, 2014.
DOI: 10.1007/978-3-662-44680-5_4

partial classification was taken into account by Ziarko in Refs. [2,3]. Ziarko's approach is referred to as the Variable Precision Rough Set (VPRS) model.

Moreover, it should be noticed that in Refs. [4–7], Qian et al. argued that we often need to describe concurrently a target concept through multi binary relations (e.g. equivalence relation, tolerance relation, reflexive relation and neighborhood relation) on the universe according to a user's requirements or targets of problem solving. Therefore, they proposed the concept of Multigranulation Rough Set (MGRS) model. The first multigranulation rough set model was proposed by Qian et al. in Ref. [5]. Following such work, Qian et al. also classified his multigranulation rough set into two parts: one is the optimistic multigranulation rough set [4,5] and the other is the pessimistic multigranulation rough set [6]. In recent years, multigranulation rough set progressing rapidly [9,13]. For example, Yang et al. and Xu et al. generalized multigranulation rough set into fuzzy environment in Ref. [10] and Ref. [11], respectively. Yang et al. [12,13] also introduced multigranulation rough sets into incomplete information system. Qian et al. [14] also developed a new multigranulation rough set decision theory, called multigranulation decision-theoretic rough sets, which combine the multigranulation idea and the Bayesian decision theory. Lin et al. [15] developed a neighborhood-based multigranulation rough set.

The purpose of this paper is to further generalize Ziarko's variable precision rough set and Qian's multigranulation rough set. From this point of view, we will propose the concept of the Variable Precision Multigranulation Rough Set (VPMGRS). On the one hand, the variable precision multigranulation rough set is a generalization of the variable precision rough set since it uses a family of the binary relations instead of single one; on the other hand, the variable precision multigranulation rough set is also a generalization of the multigranulation rough set since it takes the partial classification into account under the frame of multigranulation rough set.

To facilitate our discussion, we present the basic knowledge about variable precision rough set and multigranulation rough set in Sect. 2. In Sect. 3, we propose the model of variable precision multigranulation rough set and the basic properties of this model. In Sect. 4, we explore the relationships among the three rough set models. Since attributes reduction is one of the most important problems in rough set theory, we also apply the approximation quality significance measure to conduct the reduction in our variable precision multigranulation rough set model. Through experimental analyses, the comparisons of reducts on three different multigranulation rough sets are shown in Sect. 5. The paper ends with conclusions in Sect. 6.

2 VPRS and MGRS

2.1 Variable Precision Rough Set

Given the universe U, $\forall X, Y \subseteq U$, we say that the set X is included in the set Y with an admissible error β if and only if:

$$X \subseteq^{\beta} Y \Leftrightarrow e(X, Y) \leq \beta, \tag{1}$$

where

$$e(X,Y) = 1 - \frac{|X \cap Y|}{|X|}. \tag{2}$$

The quantity $e(X,Y)$ is referred to as the inclusion error of X in Y. The value of β should be limited: $0 \le \beta < 0.5$.

Given an information system I in which $A \subseteq AT$, the variable precision lower and upper approximations in terms of $IND(A)$ are defined as

$$\underline{A}_\beta(X) = \{x \in U : e([x]_A, X) \le \beta\}; \tag{3}$$

$$\overline{A}_\beta(X) = \{x \in U : e([x]_A, X) < 1 - \beta\}. \tag{4}$$

The pair $[\underline{A}_\beta(X), \overline{A}_\beta(X)]$ is referred to as the variable precision rough set of X with respect to the set of attributes A.

2.2 Multigranulation Rough Set

The multigranulation rough set is different from Pawlak's rough set model since the former is constructed on the basis of a family of indiscernibility relations instead of single indiscernibility relation.

In Qian's multigranulation rough set theory, two different models have been defined. The first one is the optimistic multigranulation rough set and the other is pessimistic multigranulation rough set [6].

Optimistic Multigranulation Rough Set. In Qian et al.'s optimistic multi-granulation rough set, the target is approximated through a family of the indiscernibility relations. In lower approximation, the word "optimistic" is used to express the idea that in multi independent granular structures, we need only at least one granular structure to satisfy with the inclusion condition between equivalence class and target. The upper approximation of optimistic multigranulation rough set is defined by the complement of the lower approximation. To notice what the multigranulation rough set is, we need the following notions of the information system.

Formally, an information system can be considered as a pair $I =< U, AT >$, where U is a non-empty finite set of objects, it is called the universe; AT is non-empty finite set of attributes, such that $\forall a \in AT$, V_a is the domain of attribute a.

$\forall x \in U$, let $a(x)$ denotes the value that x holds on a ($a \in AT$). For an information system I, one then can describe the relationship between objects through their attribute values. With respect to a subset of attributes such that $A \subseteq AT$, an indiscernibility relation $IND(A)$ may be defined as

$$IND(A) = \{(x,y) \in U^2 : a(x) = a(y), \forall a \in A\}. \tag{5}$$

The relation $IND(A)$ is reflexive, symmetric and transitive, then $IND(A)$ is an equivalence relation. Then $[x]_A = \{y \in U : (x,y) \in IND(A)\}$ is the A-equivalence class, which contains x.

Definition 1. *Let I be an information system in which $A_1, A_2, \ldots, A_m \subseteq AT$, then $\forall X \subseteq U$, the optimistic multigranulation lower and upper approximations are denoted by $\sum_{i=1}^m A_i{}^O(X)$ and $\overline{\sum_{i=1}^m A_i}{}^O(X)$, respectively,*

$$\sum_{i=1}^m A_i{}^O(X) = \{x \in U : [x]_{A_1} \subseteq X \vee [x]_{A_2} \subseteq X \vee \ldots [x]_{A_m} \subseteq X\}; \quad (6)$$

$$\overline{\sum_{i=1}^m A_i}{}^O(X) = \sim \sum_{i=1}^m A_i{}^O(\sim X); \quad (7)$$

where $[x]_{A_i}(1 \leq i \leq m)$ is the equivalence class of x in terms of set of attributes A_i, $\sim X$ is the complement of set X.

By the lower and upper approximations $\sum_{i=1}^m A_i{}^O(X)$ and $\overline{\sum_{i=1}^m A_i}{}^O(X)$, the optimistic multigranulation boundary region of X is

$$BN^O_{\sum_{i=1}^m A_i}(X) = \overline{\sum_{i=1}^m A_i}{}^O(X) - \sum_{i=1}^m A_i{}^O(X). \quad (8)$$

Theorem 1. *Let I be an information system in which $A_1, A_2, \ldots, A_m \subseteq AT$, then $\forall X \subseteq U$, we have*

$$\overline{\sum_{i=1}^m A_i}{}^O(X) = \{x \in U : [x]_{A_1} \cap X \neq \emptyset \wedge [x]_{A_2} \cap X \neq \emptyset \wedge \ldots$$
$$\wedge [x]_{A_m} \cap X \neq \emptyset\}. \quad (9)$$

By Theorem 1, we can see that though the optimistic multigranulation upper approximation is defined by the complement of the optimistic multigranulation lower approximation, it can also be considered as a set in which objects have non-empty intersection with the target in terms of each granular structure.

Pessimistic Multigranulation Rough Set. In Qian's pessimistic multigranulation rough set, the target is still approximated through a family of the indiscernibility relations. However, it is different from the optimistic case. In lower approximation, the word "pessimistic" is used to express the idea that in multi independent granular structures, we need all the granular structures to satisfy with the inclusion condition between equivalence class and target. The upper approximation of pessimistic multigranulation rough set is also defined by the complement of the pessimistic multigranulation lower approximation.

Definition 2. *Let I be an information system in which $A_1, A_2, \ldots, A_m \subseteq AT$, then $\forall X \subseteq U$, the pessimistic multigranulation lower and upper approximations are denoted by $\sum_{i=1}^m A_i{}^P(X)$ and $\overline{\sum_{i=1}^m A_i}{}^P(X)$, respectively,*

$$\sum_{i=1}^m A_i{}^P(X) = \{x \in U : [x]_{A_1} \subseteq X \wedge [x]_{A_2} \subseteq X \wedge \ldots [x]_{A_m} \subseteq X\}; \quad (10)$$

$$\overline{\sum_{i=1}^m A_i}{}^P(X) = \sim \sum_{i=1}^m A_i{}^P(\sim X). \quad (11)$$

By the lower and upper approximations $\underline{\sum_{i=1}^{m} A_i}^{P}(X)$ and $\overline{\sum_{i=1}^{m} A_i}^{P}(X)$, the pessimistic multigranulation boundary region of X is

$$BN_{\sum_{i=1}^{m} A_i}^{P}(X) = \overline{\sum_{i=1}^{m} A_i}^{P}(X) - \underline{\sum_{i=1}^{m} A_i}^{P}(X). \tag{12}$$

Theorem 2. *Let I be an information system in which $A_1, A_2, \ldots, A_m \subseteq AT$, then $\forall X \subseteq U$, we have*

$$\overline{\sum_{i=1}^{m} A_i}^{P}(X) = \{x \in U : [x]_{A_1} \cap X \neq \emptyset \vee [x]_{A_2} \cap X \neq \emptyset \vee \ldots$$
$$\vee [x]_{A_m} \cap X \neq \emptyset\}. \tag{13}$$

Different from the upper approximation of optimistic multigranulation rough set, the upper approximation of pessimistic multigranulation rough set is represented as a set in which objects have non-empty intersection with the target in terms of at least one granular structure.

3   Variable Precision Multigranulation Rough Set

3.1   Variable Precision Optimistic Multigranulation Rough Set

Definition 3. *Let I be an information system in which $A_1, A_2, \ldots, A_m \subseteq AT$, then $\forall X \subseteq U$, the variable precision optimistic multigranulation lower and upper approximations are denoted by $\underline{\sum_{i=1}^{m} A_i}_{\beta}^{O}(X)$ and $\overline{\sum_{i=1}^{m} A_i}_{\beta}^{O}(X)$, respectively,*

$$\underline{\sum_{i=1}^{m} A_i}_{\beta}^{O}(X) = \{x \in U : e([x]_{A_1}, X) \leq \beta \vee e([x]_{A_2}, X) \leq \beta \vee \ldots$$
$$\vee e([x]_{A_m}, X) \leq \beta\}; \tag{14}$$
$$\overline{\sum_{i=1}^{m} A_i}_{\beta}^{O}(X) = \sim \underline{\sum_{i=1}^{m} A_i}_{\beta}^{O}(\sim X). \tag{15}$$

Theorem 3. *Let I be an information system in which $A_1, A_2, \ldots, A_m \subseteq AT$, then $\forall X \subseteq U$, we have*

$$\overline{\sum_{i=1}^{m} A_i}_{\beta}^{O}(X) = \{x \in U : e([x]_{A_1}, X) < 1 - \beta \wedge e([x]_{A_2}, X) < 1 - \beta \wedge$$
$$\ldots \wedge e([x]_{A_m}, X) < 1 - \beta\}. \tag{16}$$

Theorem 4. *Let I be an information system in which $A_1, A_2, \ldots, A_m \subseteq AT$, then $\forall X, Y \subseteq U$, we have*

1. $\underline{\sum_{i=1}^{m} A_i}_{\beta}^{O}(U) = \overline{\sum_{i=1}^{m} A_i}_{\beta}^{O}(U) = U$;
2. $\underline{\sum_{i=1}^{m} A_i}_{\beta}^{O}(\emptyset) = \overline{\sum_{i=1}^{m} A_i}_{\beta}^{O}(\emptyset) = \emptyset$;
3. $\beta_1 \geq \beta_2 \Rightarrow \underline{\sum_{i=1}^{m} A_i}_{\beta_1}^{O}(X) \supseteq \underline{\sum_{i=1}^{m} A_i}_{\beta_2}^{O}(X), \overline{\sum_{i=1}^{m} A_i}_{\beta_1}^{O}(X) \subseteq \overline{\sum_{i=1}^{m} A_i}_{\beta_2}^{O}(X)$;

4. $\underline{\sum_{i=1}^{m} A_i}^{O}_{\beta}(X \cap Y) \subseteq \underline{\sum_{i=1}^{m} A_i}^{O}_{\beta}(X) \cap \underline{\sum_{i=1}^{m} A_i}^{O}_{\beta}(Y);$

5. $\underline{\sum_{i=1}^{m} A_i}^{O}_{\beta}(X \cup Y) \supseteq \underline{\sum_{i=1}^{m} A_i}^{O}_{\beta}(X) \cup \underline{\sum_{i=1}^{m} A_i}^{O}_{\beta}(Y);$

6. $\overline{\sum_{i=1}^{m} A_i}^{O}_{\beta}(X \cap Y) \subseteq \overline{\sum_{i=1}^{m} A_i}^{O}_{\beta}(X) \cap \overline{\sum_{i=1}^{m} A_i}^{O}_{\beta}(Y);$

7. $\overline{\sum_{i=1}^{m} A_i}^{O}_{\beta}(X \cup Y) \supseteq \overline{\sum_{i=1}^{m} A_i}^{O}_{\beta}(X) \cup \overline{\sum_{i=1}^{m} A_i}^{O}_{\beta}(Y).$

Proof. By Definition 3, it is not different to prove 1, 2 and 3.
$\forall x \in U,$

$$x \in \underline{\sum_{i=1}^{m} A_i}^{O}_{\beta}(X \cap Y) \Rightarrow \exists i \in \{1,\ldots,m\}, e([x]_{A_i}, X \cap Y) \leq \beta$$

$$\Rightarrow \exists i \in \{1,\ldots,m\}, 1 - \frac{|[x]_{A_i} \cap X \cap Y|}{|[x]_{A_i}|} \leq \beta$$

$$\Rightarrow \exists i \in \{1,\ldots,m\}, 1 - \frac{|[x]_{A_i} \cap X|}{|[x]_{A_i}|} \leq \beta$$

$$and \quad 1 - \frac{|[x]_{A_i} \cap Y|}{|[x]_{A_i}|} \leq \beta$$

$$\Rightarrow x \in \underline{\sum_{i=1}^{m} A_i}^{O}_{\beta}(X) \quad and \quad x \in \underline{\sum_{i=1}^{m} A_i}^{O}_{\beta}(Y)$$

$$\Rightarrow x \in \underline{\sum_{i=1}^{m} A_i}^{O}_{\beta}(X) \cap \underline{\sum_{i=1}^{m} A_i}^{O}_{\beta}(Y)$$

Similarity, it is not difficult to prove other formulas.   □

Remark 1. It should be noticed that in variable precision optimistic multigranulation rough set model, $\underline{\sum_{i=1}^{m} A_i}^{O}_{\beta}(X) \subseteq \overline{\sum_{i=1}^{m} A_i}^{O}_{\beta}(X)$ does not always hold.

3.2   Variable Precision Pessimistic Multigranulation Rough Set

Definition 4. *Let I be an information system in which $A_1, A_2, \ldots, A_m \subseteq AT$, then $\forall X \subseteq U$, the variable precision pessimistic multigranulation lower and upper approximations are denoted by $\underline{\sum_{i=1}^{m} A_i}^{P}_{\beta}(X)$ and $\overline{\sum_{i=1}^{m} A_i}^{P}_{\beta}(X)$, respectively,*

$$\underline{\sum_{i=1}^{m} A_i}^{P}_{\beta}(X) = \{x \in U : e([x]_{A_1}, X) \leq \beta \wedge e([x]_{A_2}, X) \leq \beta \wedge \ldots$$

$$\wedge e([x]_{A_m}, X) \leq \beta\}; \tag{17}$$

$$\overline{\sum_{i=1}^{m} A_i}^{P}_{\beta}(X) = \sim \underline{\sum_{i=1}^{m} A_i}^{P}_{\beta}(\sim X). \tag{18}$$

Theorem 5. *Let I be an information system in which $A_1, A_2, \ldots, A_m \subseteq AT$, then $\forall X \subseteq U$, we have*

$$\overline{\sum_{i=1}^{m} A_i}^{P}_{\beta}(X) = \{x \in U : e([x]_{A_1}, X) < 1 - \beta \vee e([x]_{A_2}, X) < 1 - \beta \vee \ldots$$

$$\vee e([x]_{A_m}, X) < 1 - \beta\}. \tag{19}$$

Theorem 6. *Let I be an information system in which $A_1, A_2, \ldots, A_m \subseteq AT$, then $\forall X, Y \subseteq U$, we have*

1. $\underline{\sum_{i=1}^m A_i}_\beta^P(U) = \overline{\sum_{i=1}^m A_i}_\beta^P(U) = U;$

2. $\underline{\sum_{i=1}^m A_i}_\beta^P(\emptyset) = \overline{\sum_{i=1}^m A_i}_\beta^P(\emptyset) = \emptyset;$

3. $\beta_1 \geq \beta_2 \Rightarrow \underline{\sum_{i=1}^m A_i}_{\beta_1}^P(X) \supseteq \underline{\sum_{i=1}^m A_i}_{\beta_2}^P(X), \overline{\sum_{i=1}^m A_i}_{\beta_1}^P(X) \subseteq \overline{\sum_{i=1}^m A_i}_{\beta_2}^P(X);$

4. $\underline{\sum_{i=1}^m A_i}_\beta^P(X \cap Y) \subseteq \underline{\sum_{i=1}^m A_i}_\beta^P(X) \cap \underline{\sum_{i=1}^m A_i}_\beta^P(Y);$

5. $\underline{\sum_{i=1}^m A_i}_\beta^P(X \cup Y) \supseteq \underline{\sum_{i=1}^m A_i}_\beta^P(X) \cup \underline{\sum_{i=1}^m A_i}_\beta^P(Y);$

6. $\overline{\sum_{i=1}^m A_i}_\beta^P(X \cap Y) \subseteq \overline{\sum_{i=1}^m A_i}_\beta^P(X) \cap \overline{\sum_{i=1}^m A_i}_\beta^P(Y);$

7. $\overline{\sum_{i=1}^m A_i}_\beta^P(X \cup Y) \supseteq \overline{\sum_{i=1}^m A_i}_\beta^P(X) \cup \overline{\sum_{i=1}^m A_i}_\beta^P(Y).$

Proof. By Definition 4, it is not difficult to prove these theorems. $\qquad\square$

Remark 2. It should be noticed that in variable precision pessimistic multigranulation rough set model, $\underline{\sum_{i=1}^m A_i}_\beta^P(X) \subseteq \overline{\sum_{i=1}^m A_i}_\beta^P(X)$ does not always hold.

3.3   Approximation Quality

As to the multigranulation case, Qian et al. have presented the definitions of approximation qualities based on optimistic and pessimistic multigranulation rough sets. In this subsection, we will generalize these measures into variable precision multigranulation rough set.

Definition 5. *Let $I = <U, AT \cup \{d\}>$ be a decision system which is introduced a set of decision attribute d into information system $<U, AT>$, $A_1, A_2, \ldots, A_m \subseteq AT$, $\{X_1, X_2, \ldots, X_n\}$ is the partition induced by set of decision attribute d, then approximation qualities of d based on variable precision optimistic and pessimistic multigranulation rough sets are defined as $\gamma^O(AT, \beta, d)$ and $\gamma^P(AT, \beta, d)$, respectively, such that*

$$\gamma^O(AT, \beta, d) = |\cup \{\underline{\sum_{i=1}^m A_i}_\beta^O(X_j) : 1 \leq j \leq n\}|/|U|; \tag{20}$$

$$\gamma^P(AT, \beta, d) = |\cup \{\underline{\sum_{i=1}^m A_i}_\beta^P(X_j) : 1 \leq j \leq n\}|/|U|. \tag{21}$$

in which $|X|$ is the cardinal number of classical set X.

Theorem 7. *Let $I = <U, AT \cup \{d\}>$ be a decision system in which $A_1, A_2, \ldots, A_m \subseteq AT$, $0 \leq \beta_1 \leq \beta_2 < 0.5$, then we have*

$$\gamma^O(AT, \beta_1, d) \leq \gamma^O(AT, \beta_2, d); \tag{22}$$

$$\gamma^P(AT, \beta_1, d) \leq \gamma^P(AT, \beta_2, d). \tag{23}$$

Proof. It can be proven directly through Theorems 4 and 6. $\qquad\square$

4 Relationships Among Several Models

Theorem 8. *Let I be an information system in which $A_1, A_2, \ldots, A_m \subseteq AT$, then $\forall X \subseteq U$, we have*

1. $\sum_{i=1}^{m} \underline{A_i}^P_\beta(X) \subseteq \sum_{i=1}^{m} \underline{A_i}^O_\beta(X)$;
2. $\sum_{i=1}^{m} \overline{A_i}^O_\beta(X) \subseteq \sum_{i=1}^{m} \overline{A_i}^P_\beta(X)$.

Proof. $\forall x \in \sum_{i=1}^{m} \underline{A_i}^P_\beta(X)$, then we have $e([x]_{A_i}, X) \leq \beta$ for each $i = 1, 2, \ldots, m$. By the definition of the variable precision optimistic multigranulation lower approximation, $x \in \sum_{i=1}^{m} \underline{A_i}^O_\beta(X)$ holds obviously, it follows that $\sum_{i=1}^{m} \underline{A_i}^P_\beta(X) \subseteq \sum_{i=1}^{m} \underline{A_i}^O_\beta(X)$.

Similarly, it is not difficult to prove $\sum_{i=1}^{m} \overline{A_i}^O_\beta(X) \subseteq \sum_{i=1}^{m} \overline{A_i}^P_\beta(X)$ through Theorems 3 and 5. $\qquad\square$

Theorem 8 shows the relationships between variable precision optimistic and pessimistic multigranulation rough approximations. The details are: the variable precision pessimistic multigranulation lower approximation is included into the variable precision optimistic multigranulation lower approximation, the variable precision optimistic multigranulation upper approximation is included into the variable precision pessimistic multigranulation upper approximation.

Theorem 9. *Let I be an information system in which $A_1, A_2, \ldots, A_m \subseteq AT$, then $\forall X \subseteq U$, we have*

1. $\sum_{i=1}^{m} \underline{A_i}^O(X) \subseteq \sum_{i=1}^{m} \underline{A_i}^O_\beta(X)$;
2. $\sum_{i=1}^{m} \overline{A_i}^O(X) \supseteq \sum_{i=1}^{m} \overline{A_i}^O_\beta(X)$;
3. $\sum_{i=1}^{m} \underline{A_i}^P(X) \subseteq \sum_{i=1}^{m} \underline{A_i}^P_\beta(X)$;
4. $\sum_{i=1}^{m} \overline{A_i}^P(X) \supseteq \sum_{i=1}^{m} \overline{A_i}^P_\beta(X)$;

Proof. $\forall x \in \sum_{i=1}^{m} \underline{A_i}^O(X)$, there $\exists i = 1, 2, \ldots, m$ such that $[x]_{A_i} \subseteq X$, i.e. $e([x]_{A_i}, X) = 0$, it follows that $e([x]_{A_i}, X) \leq \beta$, then by the definition of the variable precision optimistic multigranulation lower approximation, $x \in \sum_{i=1}^{m} \underline{A_i}^O_\beta(X)$, i.e. $\sum_{i=1}^{m} \underline{A_i}^O(X) \subseteq \sum_{i=1}^{m} \underline{A_i}^O_\beta(X)$.

Similarity, it is not difficult to prove other formulas. $\qquad\square$

Theorem 9 shows the relationships between the classical multigranulation rough set and the variable precision multigranulation rough set. The details are: the optimistic multigranulation lower approximation is included into the variable precision optimistic multigranulation lower approximation; the variable precision optimistic multigranulation upper approximation is included into the optimistic multigranulation upper approximation; the pessimistic multigranulation lower approximation is included into the variable precision pessimistic multigranulation lower approximation; the variable precision pessimistic multigranulation upper approximation is included into the pessimistic multigranulation upper approximation.

Theorem 10. *Let I be an information system in which $A_1, A_2, \ldots, A_m \subseteq AT$, then $\forall X \subseteq U$, we have*

1. $\underline{A_{i_\beta}}(X) \subseteq \sum_{i=1}^{m} \underline{A_i}_\beta^O(X), \forall i = 1, 2, \ldots, m;$
2. $\overline{\sum_{i=1}^{m} A_i}_\beta^O(X) \subseteq \overline{A_{i\beta}}(X), \forall i = 1, 2, \ldots, m;$
3. $\underline{\sum_{i=1}^{m} A_i}_\beta^P(X) \subseteq \underline{A_{i\beta}}(X), \forall i = 1, 2, \ldots, m;$
4. $\overline{A_{i\beta}}(X) \subseteq \overline{\sum_{i=1}^{m} A_i}_\beta^P(X), \forall i = 1, 2, \ldots, m.$

Proof. $\forall i = 1, 2, \ldots, m$ and $\forall x \in \underline{A_{i\beta}}(X)$, then we have $e([x]_{A_i}, X) \leq \beta$. By the definition of the variable precision optimistic multigranulation lower approximation, $x \in \sum_{i=1}^{m} \underline{A_i}_\beta^O(X)$ holds obviously, it follows that $\underline{A_{i\beta}}(X) \subseteq \sum_{i=1}^{m} \underline{A_i}_\beta^O(X)$, $\forall i = 1, 2, \ldots, m$.

Similarity, it is not difficult to prove other formulas. $\square$

Theorem 10 shows the relationships between the classical variable precision rough set and the variable precision multigranulation rough set. The details are: the classical variable precision lower approximation is included into the variable precision optimistic multigranulation lower approximation; the variable precision optimistic multigranulation upper approximation is include into the classical variable precision upper approximation; the variable precision pessimistic multigranulation lower approximation is included into the classical variable precision lower approximation; the classical variable precision upper approximation is included into the variable precision pessimistic multigranulation upper approximation.

Theorem 11. *Let I be an information system in which $A_1, A_2, \ldots, A_m \subseteq AT$, then $\forall X \subseteq U$, we have*

1. $\underline{\sum_{i=1}^{m} A_i}^P(X) \subseteq \underline{A_{i\beta}}(X), \forall i = 1, 2, \ldots, m;$
2. $\overline{\sum_{i=1}^{m} A_i}^O(X) \subseteq \overline{A_{i\beta}}(X), \forall i = 1, 2, \ldots, m.$

Proof. $\forall x \in \sum_{i=1}^{m} A_i^P(X)$, by the definition of pessimistic multigranulation rough set, we have $\forall i = 1, 2, \ldots, m$, such that $[x]_{A_i} \subseteq X$, i.e. $e([x]_{A_i}, X) = 0$, it follows that $e([x]_{A_i}, X) \leq \beta$. Then by the definition of variable precision rough set, $x \in \underline{A_{i\beta}}(X)$ holds obviously, it follows that $\underline{\sum_{i=1}^{m} A_i}^P(X) \subseteq \underline{A_{i\beta}}(X)$, $\forall i = 1, 2, \ldots, m$.

Similarity, it is not difficult to prove $\overline{\sum_{i=1}^{m} A_i}^O(X) \subseteq \overline{A_{i\beta}}(X)$. $\square$

Theorem 11 shows the relationships between the classical variable precision rough set and the classical multigranulation rough set. The details are: the pessimistic multigranulation lower approximation is included into the classical variable precision lower approximation, the optimistic multigranulation upper approximation is included into the classical variable precision upper approximation.

5   Attribute Reduction in VPMGRS

Attribute reduction plays a crucial role in the development of rough set theory. In classical rough set theory, reduct is a minimal subset of attributes, which is independent and has the same discernibility power as all of the attributes. In recent years, with respect to different requirements, different types of attribute reductions have been proposed [17–21]. In this paper, we apply the approximation quality significance measure to conduct the reduction in our variable precision multigranulation rough set model.

Definition 6. *Let $I = <U, AT \cup \{d\}>$ be a decision system, $A_1, A_2, \ldots, A_m \subseteq AT$, B is referred to as a reduct of I, which is based on variable precision optimistic multigranulation rough set, if and only if*

1. $\gamma^O(B, \beta, d) = \gamma^O(AT, \beta, d)$;
2. $\forall B' \subset B$, $\gamma^O(B', \beta, d) \neq \gamma^O(AT, \beta, d)$.

In Definition 6, we can see that a reduct is a minimal subset of AT, which preserve the approximation qualities of d based on variable precision optimistic multigranulation rough set. Similarity, it is not difficult to define the reduct based on variable precision pessimistic multigranulation rough set. To save the space of this paper, we omit it.

Let I be a decision system, suppose that $B \subseteq AT$, $\forall A_i \in B$, we define the following coefficients.

$$Sig^O_{in}(A_i, B, d) = \gamma^O(B, \beta, d) - \gamma^O(B - A_i, \beta, d); \tag{24}$$
$$Sig^P_{in}(A_i, B, d) = \gamma^P(B, \beta, d) - \gamma^P(B - A_i, \beta, d); \tag{25}$$

is the significance of A_i in B relative to decision d. $Sig^O_{in}(A_i, B, d)$ and $Sig^P_{in}(A_i, B, d)$ reflect the changes of the lower approximation qualities based on variable precision optimistic and pessimistic multigranulation rough sets, respectively, if set of attributes A_i is eliminated from AT.

Accordingly, we also define

$$Sig^O_{out}(A_i, B, d) = \gamma^O(B \cup \{A_i\}, \beta, d) - \gamma^O(B, \beta, d); \tag{26}$$
$$Sig^P_{out}(A_i, B, d) = \gamma^P(B \cup \{A_i\}, \beta, d) - \gamma^P(B, \beta, d); \tag{27}$$

where $\forall A_i \in AT - B$. $Sig^O_{out}(A_i, B, d)$ and $Sig^P_{out}(A_i, B, d)$ measure the changes of lower approximation qualities based on variable precision optimistic and pessimistic multigranulation rough sets, respectively, if set of attributes A_i is introduced into B. This measure can be used in a forward greedy attribute reduction algorithm, while $Sig_{in}^{O}(A_i, B, d)$ and $Sig_{in}^{P}(A_i, B, d)$ is applicable to determine the significance of every set of attributes in the context of approximation quality.

Formally, a forward greedy attributes reduction can be designed as follows.

Algorithm. Attributes reduction based on VPMGRS.

Input: Decision system $I = < U, AT \cup \{d\} >$.

Output: A reduct red.

Step 1: $red \leftarrow \emptyset$;

Step 2: Compute the significance for each $A_i \in AT$ with $Sig_{in}^{O}(A_i, AT, d)$;

Step 3: $red \leftarrow A_j$ where
$$Sig_{in}^{O}(A_j, AT, d) = max\{Sig_{in}^{O}(A_i, AT, d) : \forall A_i \in AT\};$$

Step 4: Do
$\quad \forall A_i \in AT - red$, compute $Sig_{out}^{O}(A_i, red, d)$;
$\quad\quad$ If $Sig_{out}^{O}(A_j, red, d) = max\{Sig_{out}^{O}(A_i, red, d) : \forall A_i \in AT - red\}$
$\quad\quad red = red \cup \{A_j\}$;
$\quad$ End
$\quad$ Until $\gamma^{O}(red, \beta, d) = \gamma^{O}(AT, \beta, d)$;

Step 5: $\forall A_i \in red$
$\quad$ If $\gamma^{O}(red - A_i, \beta, d) = \gamma^{O}(AT, \beta, d)$
$\quad\quad red = red - \{A_i\}$
$\quad$ End

Step 6: Return red;

In the above algorithm, if Sig_{in}^{O} and Sig_{out}^{O} are replaced by Sig_{in}^{P} and Sig_{out}^{P} respectively, then it can be used to compute variable precision pessimistic multigranulation lower approximation distribution reduct.

5.1 Experimental Results

In the following, through experiment analysis, we will illustrate the reducts among variable precision multigranulation rough set, classical multigranulation rough set and classical variable precision rough set. All the experiments have been carried out on a personal computer with Windows 7, Intel Core 2 DuoT5800 CPU (2.00 GHz) and 2.00 GB memory. The programming language is Matlab 2010b.

We have downloaded five public data sets from UCI repository of Machine Learning databases, which are described in Table 1. In our experiment, we assume that each attribute in a data set can induce an equivalence relation and then all attributes in a data set will induce a family of equivalence relations on the universe of discourse.

Tables 2, 3, 4, 5 and 6 show the result of reduction based on variable precision multigranulation rough set, classical precision rough set and classical multigranulation rough set, respectively.

Table 1. Data sets description.

ID	Data sets	Samples	Attributes	Decision classes
1	Dermatology	34	34	7
2	Spect Heart	267	22	2
3	Lymphography	98	18	3
4	Wdbc	569	30	4
5	Tic-Tac-Toe Endgame	958	8	2

Table 2. Reducts of variable precision optimistic multigranulation rough set

Data ID	Type of reduct	Different values of β									
		0	0.05	0.1	0.15	0.2	0.25	0.3	0.35	0.4	0.45
1	Lower	7	7	7	7	6	6	6	6	6	3
	Upper	6	21	25	17	11	14	13	6	11	4
2	Lower	2	4	12	11	11	1	1	1	1	1
	Upper	3	5	12	13	13	7	3	2	1	1
3	Lower	4	4	5	6	7	5	6	1	1	1
	Upper	3	3	4	5	5	5	7	9	9	12
4	Lower	2	3	6	6	7	5	4	1	1	1
	Upper	4	5	8	8	9	7	9	5	8	9
5	Lower	1	1	1	1	1	1	4	8	8	1
	Upper	1	1	1	1	1	1	1	1	1	1

In Theorem 7, we observe that in variable precision rough set models, the approximation qualities are changing with different values of β. Therefore, in the experiments of the reduction of variable precision multigranulation rough set, we choose different values of β to conduct the attribute reduction. Similarly, the same ideas are applied in the reduction of variable precision rough set.

By comparing with Tables 2 and 3, we can observe the following.

1. For lower approximation quality reduction, attributes in variable precision optimistic multigranulation rough set are equal or more than these of variable precision pessimistic multigranulation rough set except $\beta = 0.4$ and $\beta = 0.45$ in the 4th data set.
2. For upper approximation quality reduction, attributes in variable precision optimistic multigranulation rough set are equal or more than these of variable precision pessimistic multigranulation rough set except $\beta = 0.45$ in the 1th data set, $\beta = 0.4$ and $\beta = 0.45$ in the 2th data set, $\beta = 0.3$ and $\beta = 0.35$ in the 5th data set.

Through experimental analysis, though attributes in variable precision pessimistic multigranulation rough set may be less than these of variable precision

Table 3. Reducts of variable precision pessimistic multigranulation rough set

Data ID	Type of reduct	Different values of β									
		0	0.05	0.1	0.15	0.2	0.25	0.3	0.35	0.4	0.45
1	Lower	1	1	1	1	1	1	1	1	1	1
	Upper	1	1	1	1	1	3	6	6	6	7
2	Lower	1	1	1	5	5	6	2	1	1	1
	Upper	1	1	1	1	1	1	1	1	11	12
3	Lower	1	1	1	1	1	1	1	1	1	1
	Upper	1	1	1	1	1	1	1	6	7	5
4	Lower	1	1	1	1	1	1	1	1	8	8
	Upper	1	1	1	1	1	1	1	4	7	6
5	Lower	1	1	1	1	1	1	1	8	8	1
	Upper	1	1	1	1	1	1	8	4	1	1

Table 4. Reducts of variable precision rough set

Data ID	Type of reduct	Different values of β									
		0	0.05	0.1	0.15	0.2	0.25	0.3	0.35	0.4	0.45
1	Lower	2	2	2	2	2	2	2	2	2	2
	Upper	2	2	2	2	2	3	4	4	4	2
2	Lower	19	18	19	18	18	16	18	18	15	17
	Upper	1	1	1	1	1	1	1	1	1	1
3	Lower	5	5	5	5	6	5	5	5	2	1
	Upper	1	1	1	1	1	1	1	1	1	1
4	Lower	2	2	2	2	2	2	3	3	2	2
	Upper	2	2	2	2	2	2	2	2	2	3
5	Lower	7	7	7	7	7	7	7	7	7	1
	Upper	1	1	1	1	1	1	1	1	1	1

optimistic multigranulation rough set, the limitation of variable precision pessimistic multigranulation rough set is stricter than that of variable precision optimistic multigranulation rough set, it mean that we could obtain empty set for variable precision pessimistic multigranulation lower approximation and full universe for variable precision pessimistic multigranulation upper approximation frequently in our experiment. From this point of view, variable precision pessimistic multigranulation rough set is meaningless since we obtain nothing for certainty or uncertainty.

Table 5. Reducts of optimistic multigranulation rough set

Data ID	Attributes in lower approximate reduct	Attributes in upper approximate reduct
1	6	2
2	2	1
3	4	1
4	2	1
5	1	1

Table 6. Reducts of pessimistic multigranulation rough set

Data ID	Attributes in lower approximate reduct	Attributes in upper approximate reduct
1	1	1
2	1	1
3	1	1
4	28	1
5	1	1

Furthermore, by comparing with Tables 2 and 4, we can observe the following.

1. For lower approximation quality reduction, attributes in variable precision optimistic multigranulation rough set are equal or more than these of variable precision rough set except the 2th data set and 5th data set.
2. For upper approximation quality reduction, attributes in variable precision optimistic multigranulation rough set are equal or more than these of variable precision rough set.

Through experimental analysis, for both lower and upper approximation quality reduction, the numbers of reducts of variable precision optimistic multigranulation rough set are equal or more than these of classical variable precision rough set. Such difference is coming from the difference of requirements of variable precision optimistic multigranulation rough set and traditional variable precision rough set. In the lower approximation quality reduction case, from Theorem 10, we can observe that the classical variable precision lower approximation is included into the variable precision optimistic multigranulation lower approximation, hence, the approximation quality of classical variable precision rough set is equal to or less than the variable precision optimistic multigranulation rough set, by these analysis, we can affirm that we need more attributes in order to preserve the approximation quality. In the upper approximate quality reduction case, though the approximation quality of variable precision optimistic multigranulation is equal to or less than classical variable precision rough set and in some cases the approximation quality is zeros, but from Theorem 3, the variable precision optimistic upper approximation need all granular structures satisfy the requirement, to achieve such a goal, we also need more attributes in the reduction process.

Finally, by comparing Table 2 with Table 5 and Table 3 with Table 6, we can observe the following.

1. For lower approximation quality reduction, attributes in variable precision optimistic multigranulation rough set are equal or more than these of classical optimistic multigranulation rough set except $\beta = 0.45$ in the 1th data set and the values of β from 0.35 to 0.45 in the 3th and 4th data sets; attributes in variable precision pessimistic multigranulation rough set are equal or more than these of classical multigranulation rough set except 4th data set.
2. For upper approximation quality reduction, attributes in variable precision optimistic multigranulation rough set are equal or more than these of classical optimistic multigranulation rough set; attributes in variable precision pessimistic multigranulation rough set are equal or more than these of classical pessimistic multigranulation rough set.

Through experimental analysis, for both lower and upper approximation quality reduction, the numbers of reducts of variable precision optimistic and pessimistic multigranulation rough sets may equal or more than these of classical optimistic and pessimistic multigranulation rough sets respectively. Similarity, Such difference is coming from the difference of requirements of variable precision multigranulation rough set and classical multigranulation rough set. From Theorem 9, we can observe that the optimistic multigranulation lower approximation is included into the variable precision optimistic multigranulation lower approximation, hence, the approximation quality of classical multigranulation rough set is equal to or less than the variable precision multigranulation rough set, by these analysis, we can affirm that we need more attributes in order to preserve the approximation quality. Similar analysis can also obtained in optimistic multigranulation upper approximation quality reduction case and pessimistic case.

6   Conclusions

In this paper, the concept of the variable precision multigranulation rough set is proposed. Such model is the fusion of variable precision rough set and multigranulation rough set. Different from Ziarko's variable precision rough set, we used a family of the indiscernibility relations instead of single indiscernibility relation for constructing rough approximation. Moreover, different from Qian's multigranulation rough set, we presented the inclusion error into the multigranulation frame. Based on this framework, a general heuristic algorithm is presented to compute variable precision multigranulation rough set lower/upper approximation quality reduction. Experimental studies pertaining to five UCI data sets show the differences of approximation qualities reduction between our with variable precision rough set and classical multigranulation rough set.

Acknowledgment. This work is supported by the Natural Science Foundation of China (Nos. 61100116, 61272419, 61305058), the Natural Science Foundation of Jiangsu Province of China (Nos. BK2011492, BK2012700, BK20130471), Qing Lan Project of Jiangsu Province of China, Key Laboratory of Intelligent Perception and Systems for High-Dimensional Information (Nanjing University of Science and Technology), Ministry of Education (No. 30920130122005), Key Laboratory of Artificial Intelligence of Sichuan Province (No. 2013RYJ03), Natural Science Foundation of Jiangsu Higher Education Institutions of China (Nos. 13KJB520003, 13KJD520008), Postgraduate Innovation Foundation of University in Jiangsu Province of China under Grant No. CXLX13_707.

References

1. Pawlak, Z.: Rough Sets Theoretical Aspects of Reasoning About Data. Kluwer Academic Publishers, Dordrecht (1991)
2. Ziarko, W.: Variable precision rough set model. J. Comput. Syst. Sci. **46**, 39–59 (1993)
3. Ziarko, W.: Probabilistic approach to rough sets. Int. J. Approx. Reason. **49**, 272–284 (2008)
4. Qian, Y.H., Liang, J.Y., Yao, Y.Y., Dang, C.Y.: MGRS: a multi-granulation rough set. Inf. Sci. **180**, 949–970 (2010)
5. Qian, Y.H., Liang, J.Y., Dang, C.Y.: Incomplete multigranulation rough set. IEEE Trans. Syst. Man Cybern. **40**(2), 420–431 (2010)
6. Qian, Y.H., Liang, J.Y., Wei, W.: Pessimistic rough decision. In: Second International Workshop on Rough Sets Theory, pp. 440–449 (2010)
7. Qian, Y.H., Liang, J.Y., Witold, P., Dang, C.Y.: Positive approximation: an accelerator for attribute reduction in rough set theory. Artif. Intell. **174**, 597–618 (2010)
8. Yang, X.B., Qi, Y.S., Song, X.N., Yang, J.Y.: Test cost sensitive multigranulation rough set: model and minimal cost selection. Inf. Sci. **250**, 184–199 (2013)
9. Yang, X.B., Zhang, Y.Q., Yang, J.Y.: Local and global measurements of MGRS rules. Int. J. Comput. Intell. Syst. **5**(6), 1010–1024 (2012)
10. Xu, W.H., Wang, Q.R., Zhang, X.T.: Multi-granulation fuzzy rough set in a fuzzy tolerance approximation space. Int. J. Fuzzy Syst. **13**, 246–259 (2011)
11. Yang, X.B., Song, X.N., Chen, Z.H., Yang, J.Y.: On multigranulation rough sets in incomplete information system. Int. J. Mach. Learn. Cybern. **3**, 223–232 (2012)
12. Yang, X.B., Song, X.N., Dou, H.L., Yang, J.Y.: Multi-granulation rough set: from crisp to fuzzy case. Ann. Fuzzy Math. Inf. **1**, 55–70 (2011)
13. Yang, X.B., Yang, J.Y.: Incomplete Information System and Rough Set Theory: Models and Attribute Reductions. Science Press & Springer, Beijing and Heidelberg (2012)
14. Qian, Y.H., Zhang, H., Sang, Y.L., Liang, J.Y.: Multigranulation decision-theoretic rough sets. Int. J. Approx. Reason. **55**, 225–237 (2014)
15. Lin, G.P., Qian, Y.H., Li, J.J.: NMGRS: Neighborhood-based multigranulation rough sets. Int. J. Approx. Reason. **53**, 1080–1093 (2012)
16. Xu, W., Zhang, X., Wang, Q.: A generalized multi-granulation rough set approach. In: Huang, D.-S., Gan, Y., Premaratne, P., Han, K. (eds.) ICIC 2011. LNCS (LNBI), vol. 6840, pp. 681–689. Springer, Heidelberg (2012)
17. Dai, J.H., Wang, W.T., Xu, Q.: An uncertainty measure for incomplete decision tables and its applications. IEEE Trans. Cybern. **43**(4), 1277–1289 (2013)

18. Dai, J.H., Wang, W.T., Tian, H.W., Liu, L.: Attribute selection based on a new conditional entropy for incomplete decision systems. Knowl. Based Syst. **39**, 207–213 (2013)
19. Dai, J.H., Tian, H.W., Wang, W.T., Liu, L.: Decision rule mining using classification consistency rate. Knowl. Based Syst. **43**, 95–102 (2013)
20. Dai, J.H.: Rough set approach to incomplete numerical data. Inf. Sci. **241**, 43–57 (2013)
21. Wei, L., Li, H.R., Zhang, W.X.: Knowledge reduction based on the equivalence relations defined on attribute set and its power set. Inf. Sci. **177**, 3178–3185 (2007)

Three-Way Decisions Versus Two-Way Decisions on Filtering Spam Email

Xiuyi Jia[1,2(✉)] and Lin Shang[2]

[1] School of Computer Science and Engineering,
Nanjing University of Science and Technology, Nanjing 210094, China
jiaxy@njust.edu.cn
[2] State Key Laboratory for Novel Software Technology,
Nanjing University, Nanjing 210093, China
shanglin@nju.edu.cn

Abstract. A three-way decisions solution and a two-way decisions solution for filtering spam emails are examined in this paper. Compared to two-way decisions, the spam filtering is no longer viewed as a binary classification problem, and each incoming email is accepted as a legitimate or rejected as a spam or undecided as a further-examined email in the three-way decisions. One advantage of the three-way decisions solution for spam filtering is that it can reduce the error rate of classifying a legitimate email to spam with minimum misclassification cost. The other one is that the solution can provide a more meaningful decision procedure for users while it is not restricted to a specific classifier. Experimental results on several corpus show that the three-way decisions solution can get a lower error rate and a lower misclassification cost.

Keywords: Decision-theoretic rough set model · Spam filtering · Three-way decisions · Two-way decisions

1 Introduction

Spam filtering is an important issue for both users and mail servers. More and more spam emails are generated and sent every day. Spam is also used for phishing, spreading viruses and online fraud, causing more damage to society. For users, dealing with spam is an annoying thing, because it always costs their time and it can bring some security problems. For mail servers, the spreading of spam wastes their storage and costs their money. To solve this threat, applying spam filtering technology becomes a good option for mail service providers. A good spam filter can help users keep out of the damage from spam emails, but a bad spam filter can also bring costs. If an interview notification is labelled as a spam and dismissed by the server, the user may lose the opportunity of getting a good position. If an email with virus is passed as a legitimate and clicked, the user may angrily smash the keyboard and change the mail server.

© Springer-Verlag Berlin Heidelberg 2014
J.F. Peters et al. (Eds.): Transactions on Rough Sets XVIII, LNCS 8449, pp. 69–91, 2014.
DOI: 10.1007/978-3-662-44680-5_5

Current research on spam filtering can be categorized into two groups: one group is seeking more and more useful features for filtering. Wei et al. [21] studied the possibility of using hosting IP address to identify potential spam domains. Halder et al. [10] analyzed the usefulness of sender's name as parameter for fuzzy string matching. Benevenuto et al. [2] studied the detecting spammers on Twitter, and a number of characteristics related to tweet content and user social behavior were applied as attributes in the classification procedure. Khonji et al. [12] selected an effective feature subset to enhance the final classification accuracy of phishing email classifiers.

The other group concentrates on the classification technology. Many machine learning algorithms were employed in different filters to classify an incoming email as a legitimate email or a spam, such as Naive bayesian classifier [19], memory based classifier (k-nn) [1], SVM based classifier [7] and so on [3,20]. Khorsi [13] summarized the content-based spam filtering techniques, including Bayesian classifier, k-nn, SVM, maximum entropy model, neural networks, genetic programming and artificial immune system. DeBarr and Wechsler [5] focused on efficient construction of effective models for spam detection. Clustering messages allows for efficient labelling of a representative sample of messages for learning a spam detection model using a Random Forest for classification and active learning for refining the classification model. Fiumara et al. [9] proposed a rule-based system for end-user email annotations by considering users' participation.

The purpose of all these studies are to increase the accuracy of spam filtering. By reviewing above researches, we can find that a popular approach is to treat spam filtering as a binary classification problem. An advantage of that is many machine learning algorithms can be applied directly. When an email is coming, it will be classified into the legitimate box or the spam box with the two-way decisions made by the classifier. Then the accuracy depends on the classifier's performance mostly. For those emails with distinct spam or legitimate features, the two-way decisions method can get a good result with a simple procedure. But for those emails with both spam and legitimate features, which are not easy to recognize immediately, the "simple and crude" classification method may bring more misclassification errors.

To reduce this kind of misclassification errors, we will study a three-way decisions method to deal with spam filtering problem in this paper. Besides the usual decisions include *accept* an email as a legitimate one and *reject* an email as a spam, the third type decision *further-examined* for those suspicious spam emails is also considered in three-way decisions solution. One usually makes a decision based on available information and evidence. When the evidence is insufficient or weak, it might be impossible to make either a positive or a negative decisions [23]. The idea of three-way decisions making can be found in many areas. In [4], an optimum rejection scheme was derived to safeguard against excessive misclassification in pattern recognition system. In clinical decision making for a certain disease, with options of treating the conditional directly, not treating the condition, or performing a diagnose test to decide whether or not to treat the condition [16]. Editor in a journal often makes three kinds of decisions for a manuscript: acceptance, rejection, or further revision. Decision-theoretic rough set model (DTRS), proposed by Yao et al. [22,23], is a typical three-way decisions method, and it will be analyzed and applied in the spam filtering problem in this paper.

We have two reasons to choose decision-theoretic rough set model to deal with spam filtering problem, one is that it can provide three-way decisions handling mechanism, the other is that it is also a kind of cost sensitive learning method. It is easy to understand that spam filtering is a cost sensitive learning problem. The cost for misclassifying a legitimate email as spam far outweighs the cost of marking a spam email as legitimate [19]. Many machine learning approaches to spam filtering papers have considered the different costs of two types of mis-classifications (*legit* → *spam* and *spam* → *legit*). In [19], 99.9 % was used as the certain threshold for classifying test email as spam to reflect the asymmetric cost of errors. In [1], three scenarios($\lambda = 1$, $\lambda = 9$, $\lambda = 999$) were discussed, while *legit* → *spam* is λ times costly than *spam* → *legit*. In these papers, the parameters were applied in the cost sensitive evaluation procedure only, which can measure the efficiency of the learning algorithm, but can not help learn a better result because they were not considered in the training procedure. In these papers, the thresholds were set manually, and appropriate settings of parameters were not discussed. Fortunately, decision-theoretic rough set model provides a semantic explanation and systematically computation of required thresholds.

More recently, Zhao et al. [24] introduced an email classification schema based on decision-theoretic rough set model by classifying the incoming email into three categories. Zhou et al. [25] proposed a three-way decision approach to email spam filtering by combining decision-theoretic rough set model and Naive Bayesian classifier. Perez-Diaz et al. [18] focused on dealing with *suspicious* or *ambiguous* emails to get the final binary classification result. Our work is the continuation of them. An important conclusion induced from our work is that the three-way decisions solution is a useful decision framework by applying many different classifiers.

The main advantage of three-way decisions solution is that it allows the possibility of refusing to make a direct decision, which means it can convert some potential misclassifications into rejections, and these emails will be further-examined by users. Several cost functions are defined to state how costly each decision is, and the final decision can make the overall cost minimal filtering. We apply the three-way decisions solution on several classical classifiers which are usually used in spam filtering, including Naive Bayesian classifier, *k*-nn classifier and C4.5 classifier. The tests on several benchmark corpus[1] show the efficiency of the three-way decisions solution. We can get a lower error rate and a lower misclassification cost.

The rest of the paper is organized as follows. In Sect. 2, we review the main ideas of decision-theoretic rough set model. In Sect. 3, we introduce the three-way decisions and two-way decisions solutions on spam filtering problem with

---

[1] All corpus are available from http://labs-repos.iit.demokritos.gr/skel/i-config/.

several evaluation criteria. Section 4 shows the efficiency of the three-way decisions solution. Some concluding remarks are given in Sect. 5.

2 Decision-Theoretic Rough Set Model

Decision-theoretic rough set model was proposed by Yao et al. [22], which is based
on Bayesian decision theory. The basic ideas of the theory [23] are reviewed.

Definition 1. *A decision table is the following tuple:*

$$S = (U, At = C \cup D, \{V_a | a \in At\}, \{I_a | a \in At\}), \tag{1}$$

*where U is a finite nonempty set of objects, At is a finite nonempty set of
attributes, C is a set of condition attributes describing the objects, and D is
a set of decision attributes that indicates the classes of objects. V_a is a nonempty
set of values of $a \in At$, and $I_a : U \to V_a$ is an information function that maps
an object in U to exactly one value in V_a.*

In a decision table, an object x is described by its equivalence class under a set
of attributes $A \subseteq At$: $[x]_A = \{y \in U | \forall a \in A(I_a(x) = I_a(y))\}$. Let π_A denote the
partition induced by the set of attributes $A \subseteq At$ and $\pi_D = \{D_1, D_2, \dots, D_m\}$
denote the partition of the universe U induced by the set of decision attributes D.

Let $\Omega = \{\omega_1, \dots, \omega_s\}$ be a finite set of s states and let $\mathcal{A} = \{a_1, \dots, a_m\}$
be a finite set of m possible actions. Let $\lambda(a_i | \omega_j)$ denote the cost, for taking
action a_i when the state is ω_j. Let $p(\omega_j | x)$ be the conditional probability of an
incoming email x being in state ω_j, suppose action a_i is taken. The expected
cost associated with taking action a_i is given by:

$$R(a_i | x) = \sum_{j=1}^{s} \lambda(a_i | \omega_j) \cdot p(\omega_j | x). \tag{2}$$

In rough set theory [17], a set C is approximated by three regions, namely,
the positive region $\mathrm{POS}(C)$ includes the objects that are sure belong to C, the
boundary region $\mathrm{BND}(C)$ includes the objects that are possible belong to C, and
the negative region $\mathrm{NEG}(C)$ includes the objects that are not belong to C. In
spam filtering, we have a set of two states $\Omega = \{C, C^c\}$ indicating that an email
is in C (i.e., legitimate) or not in C (i.e., spam), respectively. The set of emails
can be divided into three regions, $\mathrm{POS}(C)$ includes emails that are legitimate
emails, $\mathrm{BND}(C)$ includes emails that need further-examination, and $\mathrm{NEG}(C)$
includes emails that are spam. With respect to these three regions, the set of
actions is given by $\mathcal{A} = \{a_P, a_B, a_N\}$, where a_P, a_B and a_N represent the three
actions in classifying an email x, namely, deciding $x \in \mathrm{POS}(C)$, deciding $x \in
\mathrm{BND}(C)$, and deciding $x \in \mathrm{NEG}(C)$. Six cost functions are imported, λ_{PP}, λ_{BP}
and λ_{NP} denote the costs incurred for taking actions a_P, a_B, a_N, respectively,
when an email belongs to C, and λ_{PN}, λ_{BN} and λ_{NN} denote the costs incurred
for taking these actions when the email does not belong to C. The cost functions
regarding the cost of actions in different states is given by the 3×2 matrix:

Table 1. Cost matrix in decision-theoretic rough set model

	$C(P)$	$C^c(N)$
P	λ_{PP}	λ_{PN}
B	λ_{BP}	λ_{BN}
N	λ_{NP}	λ_{NN}

The expected costs associated with taking different actions for email x can be expressed as:

$$R(a_P|x) = \lambda_{PP} \cdot p(C|x) + \lambda_{PN} \cdot p(C^c|x),$$
$$R(a_B|x) = \lambda_{BP} \cdot p(C|x) + \lambda_{BN} \cdot p(C^c|x),$$
$$R(a_N|x) = \lambda_{NP} \cdot p(C|x) + \lambda_{NN} \cdot p(C^c|x). \tag{3}$$

The Bayesian decision procedure suggests the following minimum-cost decision rules:

(P) If $R(a_P|x) \le R(a_B|x)$ and $R(a_P|x) \le R(a_N|x)$, decide $x \in \text{POS}(C)$;
(B) If $R(a_B|x) \le R(a_P|x)$ and $R(a_B|x) \le R(a_N|x)$, decide $x \in \text{BND}(C)$;
(N) If $R(a_N|x) \le R(a_P|x)$ and $R(a_N|x) \le R(a_B|x)$, decide $x \in \text{NEG}(C)$;

Consider a special kind of cost functions with:

$$\lambda_{PP} \le \lambda_{BP} < \lambda_{NP},$$
$$\lambda_{NN} \le \lambda_{BN} < \lambda_{PN}. \tag{4}$$

That is, the cost of classifying an email x being in C into the positive region $\text{POS}(C)$ is less than or equal to the cost of classifying x into the boundary region $\text{BND}(C)$, and both of these costs are strictly less than the cost of classifying x into the negative region $\text{NEG}(C)$. The reverse order of costs is used for classifying an email not in C. Since $p(C|x) + p(C^c|x) = 1$, under above condition, we can simplify decision rules (P)-(N) as follows:

(P) If $p(C|x) \ge \alpha$ and $p(C|x) \ge \gamma$, decide $x \in \text{POS}(C)$;
(B) If $p(C|x) \le \alpha$ and $p(C|x) \ge \beta$, decide $x \in \text{BND}(C)$;
(N) If $p(C|x) \le \beta$ and $p(C|x) \le \gamma$, decide $x \in \text{NEG}(C)$.

Where

$$\alpha = \frac{(\lambda_{PN} - \lambda_{BN})}{(\lambda_{PN} - \lambda_{BN}) + (\lambda_{BP} - \lambda_{PP})},$$
$$\beta = \frac{(\lambda_{BN} - \lambda_{NN})}{(\lambda_{BN} - \lambda_{NN}) + (\lambda_{NP} - \lambda_{BP})},$$
$$\gamma = \frac{(\lambda_{PN} - \lambda_{NN})}{(\lambda_{PN} - \lambda_{NN}) + (\lambda_{NP} - \lambda_{PP})}. \tag{5}$$

Each rule is defined by two out of the three parameters. The conditions of rule (B) suggest that $\alpha > \beta$ may be a reasonable constraint; it will ensure a well-defined boundary region. If we obtain the following condition on the cost functions [23]:

$$\frac{(\lambda_{NP} - \lambda_{BP})}{(\lambda_{BN} - \lambda_{NN})} > \frac{\lambda_{BP} - \lambda_{PP}}{(\lambda_{PN} - \lambda_{BN})}, \tag{6}$$

$0 \le \beta < \gamma < \alpha \le 1$. In this case, after tie-breaking, the following simplified rules are obtained:

(P1) If $p(C|x) \ge \alpha$, decide $x \in POS(C)$;
(B1) If $\beta < p(C|x) < \alpha$, decide $x \in BND(C)$;
(N1) If $p(C|x) \le \beta$, decide $x \in NEG(C)$.

The threshold parameters can be systematically calculated from cost functions based on the Bayesian decision theory.

3   Three-Way Decisions and Two-Way Decisions Solutions on Spam Filtering

3.1   Three-Way Decisions Solution

Given the cost functions, we can make the proper decisions for incoming emails based on the parameters (α, β), which are computed by cost functions, and the probability of each email being a legitimate one, which is provided by the running classifier.

For an email x, if the probability of being a legitimate email is $p(C|x)$, considering the special case $(\alpha > \beta)$, the three-way decisions solution is:

If $p(C|x) > \alpha$, then x is a legitimate email;
If $p(C|x) < \beta$, then x is a spam;
If $\beta \le p(C|x) \le \alpha$, then x needs further-examined.

3.2   Two-Way Decisions Solution

We will examine two kinds of two-way decisions solutions. For the first solution, spam filtering is treated as a classical classification problem. Assume the probability of being a legitimate email is $p(C|x)$ gotten from a particular classifier, then the two-way decisions solution is obey to the majority rule:

If $p(C|x) \ge 0.5$, then x is a legitimate email;
If $p(C|x) < 0.5$, then x is a spam.

0.5 is used as the threshold because of $p(C|x) + p(C^c|x) = 1$.

For the second solution, spam filtering is seen as a cost sensitive learning problem, then the method is also a cost sensitive learning two-way decisions solution. Similarly with the cost matrix in decision-theoretic rough set model, the cost matrix in two way decisions solution is given by the 2×2 matrix in Table 2. λ_{PP} and λ_{NP} denote the costs incurred for classifying a legitimate email

Table 2. Cost matrix in two-way decisions solution

	$C(P)$	$C^c(N)$
P	λ_{PP}	λ_{PN}
N	λ_{NP}	λ_{NN}

as a legitimate email and a spam, respectively. λ_{PN} and λ_{NN} denote the costs incurred for classifying a spam as a legitimate email and a spam, respectively.

Consider a common case where we assume zero cost for a correct classification, namely, $\lambda_{PP} = \lambda_{NN} = 0$, then the cost sensitive learning two-way decisions solution is defined as:

If $p(C|x) \geq \frac{\lambda_{PN}}{\lambda_{NP}+\lambda_{PN}}$, then x is a legitimate email;

If $p(C|x) < \frac{\lambda_{PN}}{\lambda_{NP}+\lambda_{PN}}$, then x is a spam.

3.3   Evaluation Measures

In this paper, two groups of measures will be checked in the experiments to compare three-way decisions and two-way decisions solutions.

Measure Group 1. As a cost sensitive learning problem, Androutsopoulos et al. [1] suggested using weighted accuracy (or weighted error rate) and total cost ratio to measure the spam filter performance. Let N_{legit} and N_{spam} be the total numbers of legitimate and spam emails, to be classified by the filter, and $n_{legit \to spam}$ the number of emails belonging to legitimate class that the filter classified as belonging to spam, $n_{spam \to legit}$, reversely.

If $legit \to spam$ is λ times more costly than $spam \to legit$, then weighted accuracy ($WAcc$) and weighted error rate ($WErr$) are defined as:

$$WAcc = \frac{\lambda \cdot n_{legit \to legit} + n_{spam \to spam}}{\lambda \cdot N_{legit} + N_{spam}}, \tag{7}$$

$$WErr = \frac{\lambda \cdot n_{legit \to spam} + n_{spam \to legit}}{\lambda \cdot N_{legit} + N_{spam}}. \tag{8}$$

As the values of accuracy and error rate (or their weighted versions) are often misleadingly high, another measure is defined to get a clear picture of a classifier's performance, the ratio of its error rate and that of a simplistic baseline approach. The baseline approach is the filter that never blocks legitimate emails and always passes spam emails. The weighted error rate of the baseline is:

$$WErr^b = \frac{N_{spam}}{\lambda \cdot N_{legit} + N_{spam}}. \tag{9}$$

The total cost ratio (TCR) is:

$$TCR = \frac{WErr^b}{WErr} = \frac{N_{spam}}{\lambda \cdot n_{legit \to spam} + n_{spam \to legit}}. \tag{10}$$

Greater TCR values indicate better performance. For $TCR < 1$, the baseline is better. If cost is proportional to wasted time, an intuitive meaning for TCR is the following: it measures how much time is wasted to delete manually all spam emails when no filter is used, compared to the time wasted to delete manually any spam emails that passed the filter plus the time needed to recover from mistakenly blocked legitimate emails.

For our three-way decisions solution, weighted rejection rate ($WRej$) is defined to indicate the weighted ratio of emails need further-exam.

$$WRej = \frac{\lambda \cdot n_{legit \rightarrow boundary} + n_{spam \rightarrow boundary}}{\lambda \cdot N_{legit} + N_{spam}}, \tag{11}$$

while $n_{legit \rightarrow boundary}$ and $n_{spam \rightarrow boundary}$ mean the numbers of legitimate and spam emails being classified to boundary (need further-exam), $WRej = 1 - WAcc - WErr$, actually. For classical classifiers, $WRej = 0$.

Measure Group 2. In this group, we will test usual measures to compare three-way decisions and two-way decisions solutions, while measure group 1 can be seen as a kind of customized evaluation measure for spam filtering. The classical accuracy (or error rate) and misclassification cost will be adopted.

The classical accuracy and error rate are defined as:

$$Acc = \frac{n_{legit \rightarrow legit} + n_{spam \rightarrow spam}}{N_{legit} + N_{spam}}, \tag{12}$$

$$Err = \frac{n_{legit \rightarrow spam} + n_{spam \rightarrow legit}}{N_{legit} + N_{spam}}. \tag{13}$$

On the basis of cost matrix (Table 1) defined in three-way decisions solution and cost matrix (Table 2) defined in two-way decisions solution, the misclassification cost for three-way decisions solution and two-way decisions solution can be defined with considering the cost functions. For the misclassification cost of two-way decisions solution, the cost contains two parts: the cost of $legit \rightarrow spam$ and the cost of $spam \rightarrow legit$:

$$mc_2 = \lambda_{NP} \cdot n_{legit \rightarrow spam} + \lambda_{PN} \cdot n_{spam \rightarrow legit}. \tag{14}$$

For the misclassification cost of three-way decisions solution, the cost consists of four parts: the cost of $legit \rightarrow spam$, the cost of $legit \rightarrow boundary$, the cost of $spam \rightarrow boundary$ and the cost of $spam \rightarrow legit$:

$$mc_3 = \lambda_{NP} \cdot n_{legit \rightarrow spam} + \lambda_{BP} \cdot n_{legit \rightarrow boundary}$$
$$+ \lambda_{BN} \cdot n_{spam \rightarrow boundary} + \lambda_{PN} \cdot n_{spam \rightarrow legit}. \tag{15}$$

4   Experiments

In this section, we will check the efficiency of the three-way decisions solution. There are two groups of experiments, one is testing measure group 1 and the second one is testing measure group 2.

4.1 Experiment 1 with Measure Group 1

The followings are the detail of classifiers, corpus and thresholds used in the first experiment.

Classifiers. Three classical classifiers (Naive Bayesian classifier, k-nn classifier and SVM classifier) are applied in the experiments.

The Naive Bayesian classifier (NB) can provide the probability of an incoming email being a legitimate email. Suppose an email x is described by a feature vector $\mathbf{x} = (\mathbf{x}_1, \mathbf{x}_2, \ldots, \mathbf{x}_n)$, where $\mathbf{x}_1, \mathbf{x}_2, \ldots, \mathbf{x}_n$ are the values of attributes of the email. Let C denote the legitimate class. Based on Bayes' theorem and the theorem of total probability, given the vector of an email, the probability of being a legitimate one is:

$$p(C|x) = \frac{p(C) \cdot p(x|C)}{p(x)}, \tag{16}$$

where $p(x) = p(x|C) \cdot p(C) + p(x|C^c) \cdot p(C^c)$. Here $p(C)$ is the prior probability of an email being in the legitimate class. $p(C)$ is commonly known as the likelihood of an email being in the legitimate class with respect to x. The likelihood $p(x|C)$ is a joint probability of $p(\mathbf{x}_1, \mathbf{x}_2, \ldots, \mathbf{x}_n|C)$.

The k-nearest neighbors algorithm (k-nn) classifier [15] predicts an email's class by a majority vote of its neighbors. Euclidean distance is usually used as the distance metric. Let $n(l)$ denote the legitimate emails' number in the k nearest neighbors, and $n(s)$ denote the spam's number. So $n(l) + n(s) = k$, then we use sigmoid function to estimate the posterior probability of an email x being in the legitimate class.

$$p(C|x) = \frac{1}{1 + \exp(-a \cdot (n(l) - n(s)))}. \tag{17}$$

In our experiments, $a = 1$ and $k = 17$.

For SVM classifier, the sigmoid function also be used to estimate the posterior probability:

$$p(C|x) = \frac{1}{1 + \exp(-a \cdot d(x))}. \tag{18}$$

While $d(x) = \frac{\mathbf{w} \cdot x + b}{\|\mathbf{w}\|}$, where weighted vector $\mathbf{w}$ and threshold b are used to define the hyperplane. $a = 6$ in our experiments.

Benchmark Corpus. Three benchmark corpus are used in this paper, called Ling-Spam, PU-Corpora and Enron-Spam.

The corpus Ling-Spam was preprocessed as three different type of corpus, named Ling-Spam-bare, Ling-Spam-lemm and Ling-Spam-stop, respectively. Ling-Spam-bare is a kind of "words only" dataset, each attribute shows if a particular word occurs in the email. For Ling-Spam-lemm, a lemmatizer was applied to Ling-Spam, which means each word was substituted by its base form(e.g. "earning" becomes "earn"). Ling-Spam-stop is a data set generated by considering stop-list based on Ling-Spam-lemm.

The corpus PU-Corpora contains PU1, PU2, PU3 and PUA corpora. All corpus are only in "bare" form: tokens were separated by white characters, but no lemmtizer or stop-list has been applied.

Another corpus Enron-Spam is divided to 6 datasets, each dataset contains legitimate emails from a single user of the Enron corpus, to which fresh spam emails with varying legit-spam ratios were added.

Since the corpus in Ling-Spam and PU-Corpora were divided into 10 datasets, 10-fold cross-validation is applied in these corpus. Enron-Spam was divided into 6 datasets, 6-fold cross-validation is applied in it's experiment. Because we just want to compare the three-way decisions solution with the two-way decisions solution based on the same classifiers and the same corpus, then it is non-necessary to apply feature selection procedures in corpus, all attributes are used in the experiments.

Experimental Result. In our experiments, three different λ values($\lambda = 1$, $\lambda = 3$, and $\lambda = 9$) are applied, same values were considered in [25]. For three-way decisions solution, we set $\alpha = 0.9$ and $\beta = 0.1$ to make corresponding decisions.

From the results showed in Tables (A1–A8), we can see that the values of *WAcc* in three-way decisions solution are lower than that in classical approaches. It is the result of moving some emails into the boundary for further-examination. We can also conclude that the three-way decisions solution decreases the values of *WErr* and increases the values of *TCR* from the results, which means the three-way decisions solution can reduce the number of "wrong classified" emails. For those "suspicious spam" emails, it is a better choice to let users decide which is a legitimate email or a spam. The increment of *TCR* shows that the three-way decisions solution gives a better performance. There exists a kind of tradeoff between the error rate and the rejection rate. Users can decrease the error rate by increasing the rejection rate.

The most important conclusion we can get from the experiments result is that, the three-way decisions solution can get a better performance than the two-way decisions solution under the same situation, it does not depend on the parameter λ or some special classifiers.

4.2   Experiment 2 with Measure Group 2

Experiment Setting. In this experiment, three classifiers are also examined including Naive Bayesian classifier (NB), k-nn classifier and C4.5 classifier. All classifiers are implemented in WEKA [11]. For the data sets, Ling-Spam-stop, PU1, PU2 and PUA corpus are selected. 10-fold cross-validation is also applied in these corpus. We shall abbreviate the three-way decisions solution based on all three classifiers to "SOLU-THREE", the two-way decisions solution to "SOLU-TWO" and the cost sensitive two-way decisions solution to "SOLU-TWO-COST" in the result tables.

Four groups of cost functions detailed in Table A9 are examined. In these four groups, all cost functions are normalized.

– For Group 1, $\lambda_{NP} = \lambda_{PN}$ and $\alpha = \beta = 0.5$ is a special case for three-way decisions solution, which means only those emails with probability 0.5 need further-examination. In this situation, three-way decisions solution nearly equals to two-way decisions solution.

– For Group 2, $\lambda_{NP} \gg \lambda_{PN}$, $\alpha = 0.5294$ and $\beta = 0.0108$ means that classifying a legitimate email as a spam brings higher cost than classifying a spam as a legitimate email. In this situation, more emails will be passed by the filter as legitimate emails. This is a kind of common used group of thresholds in the spam filtering problem. For most users, they would waste time to delete a spam rather than missing some important emails.

– For Group 3, $\lambda_{NP} = \lambda_{PN}$, $\alpha = 0.9252$ and $\beta = 0.0108$ means that both classifying a legitimate mail as a spam and classifying a spam as a legitimate email have the same cost values, and the cost value is far higher than that of classifying an email into the boundary for further-examination. In this situation, more emails will be classified into the boundary region.

– For Group 4, $\lambda_{NP} \ll \lambda_{PN}$, $\alpha = 0.9252$ and $\beta = 0.5$ means that classifying a spam as a legitimate email brings higher cost than classifying a legitimate email as a spam. In this situation, more emails will be classified as spam emails.

– For all groups, the cost of misclassification is higher than that of classifying suspicious email into the boundary region.

Experiment Results. Tables (A10–A13) show the accuracy and error rate of different classifiers on different data sets. Tables (A14–A17) show the misclassification cost of different classifiers on different data sets. From these tables, we can find some results as following:

+ Three-way decisions solution can get the least error rate based on all classifiers, with four groups of thresholds on all data sets, except one value (Naive Bayesian classifier on PU2 with Group 4, in Table A12).

+ Two-way decisions solution can get the best result on accuracy as it is a suitable measure for the classical binary classification method. Three-way decisions solution generates the boundary region for suspicious emails, which makes lower error rate and lower accuracy. But three-way decisions solution is not the worse one on the accuracy measure, cost sensitive two-way decisions solution gets the lowest accuracy on several values in the experiment results.

+ In most results, three-way decisions solution can get the least misclassification cost.

+ Independent from the thresholds setting, by comparing the classical accuracy and error rate, C4.5 is the best classifier for Ling-Spam-stop and PU2, and Naive Bayesian is the best classifier for PU1 and PUA.

+ For Group 2, as a most common used thresholds setting in the spam filtering problem, three-way decisions solution gets the best result on error rate and misclassification cost measures, except one value (the misclassification cost of k-nn classifier on PUA, in Table A17).

+ For misclassification cost measure, one unexpected interesting result is that the classical two-way decisions solution is better than cost sensitive learning two-way decisions solution. We will do more analysis and experiments in our further work to check this point.

5   Conclusions

In this paper, spam filtering is seen as a cost sensitive learning problem. From the point of optimum decision view, three-way decisions solution is a reasonable and efficient approach to spam filtering. For those suspicious emails, the three-way decisions solution moves them to a boundary region for further-examination, which can convert potential misclassification into rejections. The solution can be applied in any classifier only if the classifier can provide the probability of each mail being a legitimate. Naive Bayesian classifier, k-nn, C4.5 and SVM classifiers by considering three-way decisions solution are examined on several corpus, and the results of two groups of evaluation measures show that three-way decisions can get a better performance on the spam filtering problem. With the three-way decisions solution, we can get a lower error rate and a lower misclassification cost.

Acknowledgments. This research is supported by the National Natural Science Foundation of China under Grant No. 61170180, and the China Postdoctoral Science Foundation under Grant No. 2013M530259, and Postdoctoral Science Foundation of Jiangsu Province under Grant No. 1202021C and Natural Science Foundation of Jiangsu Province under Grant No. BK20140800.

A   Experimental Result Tables

Table A1. Measure group 1 results on corpora Ling-Spam-bare.

NB		WAcc	WErr	WRej	TCR
$\lambda = 1$	NB	0.8989	0.1011	0.0000	1.6808
	Three-way	0.8646	0.0693	0.0661	2.5115
$\lambda = 3$	NB	0.9512	0.0488	0.0000	1.3232
	Three-way	0.9251	0.0320	0.0430	2.1034
$\lambda = 9$	NB	0.9716	0.0284	0.0000	0.8488
	Three-way	0.9487	0.0174	0.0339	1.6006
k-nn					
$\lambda = 1$	k-nn	0.8832	0.1168	0.0000	1.4963
	Three-way	0.7863	0.0710	0.1427	2.9550
$\lambda = 3$	k-nn	0.9320	0.0680	0.0000	1.0979
	Three-way	0.8419	0.0321	0.1260	2.2840
$\lambda = 9$	k-nn	0.9511	0.0489	0.0000	0.7853
	Three-way	0.8636	0.0169	0.1195	1.7052

Table A1. (*Continued*)

SVM		WAcc	WErr	WRej	TCR
$\lambda = 1$	SVM	0.9393	0.0607	0.0000	9.8972
	Three-way	0.8886	0.0272	0.0843	23.0962
$\lambda = 3$	SVM	0.9371	0.0629	0.0000	6.4686
	Three-way	0.8932	0.0279	0.0790	13.7574
$\lambda = 9$	SVM	0.9321	0.0679	0.0000	4.4352
	Three-way	0.8920	0.0300	0.0780	8.5106

Table A2. Measure group 1 results on corpora Ling-Spam-lemm.

NB		WAcc	WErr	WRej	TCR
$\lambda = 1$	NB	0.9049	0.0951	0.0000	1.7984
	Three-way	0.8706	0.0644	0.0651	2.6775
$\lambda = 3$	NB	0.9529	0.0471	0.0000	1.3821
	Three-way	0.9289	0.0301	0.0410	2.1521
$\lambda = 9$	NB	0.9717	0.0283	0.0000	0.8479
	Three-way	0.9517	0.0167	0.0316	1.4485
k-nn					
$\lambda = 1$	k-nn	0.8889	0.1111	0.0000	1.5702
	Three-way	0.7955	0.0696	0.1349	2.9454
$\lambda = 3$	k-nn	0.9375	0.0625	0.0000	1.1724
	Three-way	0.8532	0.0308	0.1160	2.3681
$\lambda = 9$	k-nn	0.9566	0.0434	0.0000	0.8375
	Three-way	0.8758	0.0156	0.1086	1.8028
SVM					
$\lambda = 1$	SVM	0.9332	0.0668	0.0000	2.6733
	Three-way	0.9000	0.0408	0.0592	4.2062
$\lambda = 3$	SVM	0.9591	0.0409	0.0000	1.7334
	Three-way	0.9350	0.0226	0.0424	2.9567
$\lambda = 9$	SVM	0.9693	0.0307	0.0000	0.8859
	Three-way	0.9487	0.0154	0.0359	1.6749

Table A3. Measure group 1 results on corpora Ling-Spam-stop.

NB		WAcc	WErr	WRej	TCR
$\lambda = 1$	NB	0.8869	0.1131	0.0000	1.5081
	Three-way	0.8426	0.0782	0.0793	2.2700
$\lambda = 3$	NB	0.9430	0.0570	0.0000	1.1375
	Three-way	0.9124	0.0350	0.0526	1.9059
$\lambda = 9$	NB	0.9650	0.0350	0.0000	0.6772
	Three-way	0.9398	0.0181	0.0421	1.3371
k-nn					
$\lambda = 1$	k-nn	0.8338	0.1662	0.0000	1.1319
	Three-way	0.6543	0.0568	0.2890	7.8151
$\lambda = 3$	k-nn	0.8374	0.1626	0.0000	0.5588
	Three-way	0.6744	0.0537	0.2719	4.9959
$\lambda = 9$	k-nn	0.8389	0.1611	0.0000	0.2683
	Three-way	0.6823	0.0525	0.2652	3.4376
SVM					
$\lambda = 1$	SVM	0.9284	0.0716	0.0000	2.4032
	Three-way	0.8861	0.0446	0.0692	3.9029
$\lambda = 3$	SVM	0.9584	0.0417	0.0000	1.6406
	Three-way	0.9269	0.0217	0.0514	2.9681
$\lambda = 9$	SVM	0.9701	0.0299	0.0000	0.8998
	Three-way	0.9429	0.0127	0.0444	1.9990

Table A4. Measure group 1 results on corpora PU1

NB		WAcc	WErr	WRej	TCR
$\lambda = 1$	NB	0.9174	0.0826	0.0000	5.9340
	Three-way	0.9110	0.0761	0.0128	6.4019
$\lambda = 3$	NB	0.9558	0.0442	0.0000	5.3140
	Three-way	0.9519	0.0411	0.0069	5.7487
$\lambda = 9$	NB	0.9769	0.0231	0.0000	4.4430
	Three-way	0.9744	0.0219	0.0037	4.8404
k-nn					
$\lambda = 1$	k-nn	0.8807	0.1193	0.0000	4.1672
	Three-way	0.7661	0.0560	0.1780	8.9867

Table A4. (*Continued*)

k-nn		WAcc	WErr	WRej	TCR
$\lambda = 3$	k-nn	0.8727	0.1273	0.0000	2.0298
	Three-way	0.7468	0.0654	0.1879	3.9120
$\lambda = 9$	k-nn	0.8683	0.1317	0.0000	0.8361
	Three-way	0.7362	0.0705	0.1933	1.5428
SVM					
$\lambda = 1$	SVM	0.9789	0.0211	0.0000	22.4000
	Three-way	0.9248	0.0055	0.0697	40.0000
$\lambda = 3$	SVM	0.9797	0.0203	0.0000	13.2656
	Three-way	0.9342	0.0043	0.0615	24.0000
$\lambda = 9$	SVM	0.9801	0.0199	0.0000	9.0223
	Three-way	0.9394	0.0037	0.0570	18.6667

Table A5. Measure group 1 results on corpora PU2.

NB		WAcc	WErr	WRej	TCR
$\lambda = 1$	NB	0.8423	0.1577	0.0000	1.3128
	Three-way	0.8408	0.1577	0.0014	1.3128
$\lambda = 3$	NB	0.9135	0.0865	0.0000	1.0048
	Three-way	0.9130	0.0865	0.0005	1.0048
$\lambda = 9$	NB	0.9423	0.0577	0.0000	0.6925
	Three-way	0.9421	0.0577	0.0002	0.6925
k-nn					
$\lambda = 1$	k-nn	0.8282	0.1718	0.0000	1.1586
	Three-way	0.8239	0.1648	0.0113	1.2088
$\lambda = 3$	k-nn	0.9341	0.0659	0.0000	1.1586
	Three-way	0.9324	0.0632	0.0043	1.2088
$\lambda = 9$	k-nn	0.9769	0.0231	0.0000	1.1586
	Three-way	0.9763	0.0222	0.0015	1.2088
SVM					
$\lambda = 1$	SVM	0.9521	0.0479	0.0000	6.1600
	Three-way	0.9028	0.0169	0.0803	10.3333
$\lambda = 3$	SVM	0.9697	0.0303	0.0000	4.1600
	Three-way	0.9335	0.0108	0.0557	9.1333
$\lambda = 9$	SVM	0.9768	0.0232	0.0000	3.2484
	Three-way	0.9459	0.0083	0.0457	8.4747

Table A6. Measure group 1 results on corpora PU3.

NB		WAcc	WErr	WRej	TCR
$\lambda = 1$	NB	0.9138	0.0862	0.0000	5.6911
	Three-way	0.9107	0.0799	0.0094	6.1882
$\lambda = 3$	NB	0.9465	0.0535	0.0000	4.7892
	Three-way	0.9446	0.0494	0.0061	5.2705
$\lambda = 9$	NB	0.9644	0.0356	0.0000	3.6175
	Three-way	0.9632	0.0326	0.0042	4.0650
k-nn					
$\lambda = 1$	k-nn	0.9048	0.0952	0.0000	4.8235
	Three-way	0.8320	0.0484	0.1196	9.9110
$\lambda = 3$	k-nn	0.9359	0.0641	0.0000	3.4780
	Three-way	0.8635	0.0336	0.1029	6.8689
$\lambda = 9$	k-nn	0.9529	0.0471	0.0000	2.0081
	Three-way	0.8808	0.0255	0.0937	4.0087
SVM					
$\lambda = 1$	SVM	0.9688	0.0312	0.0000	17.1363
	Three-way	0.9334	0.0157	0.0508	37.9751
$\lambda = 3$	SVM	0.9702	0.0298	0.0000	10.1460
	Three-way	0.9359	0.0145	0.0496	26.0547
$\lambda = 9$	SVM	0.9709	0.0291	0.0000	4.9085
	Three-way	0.9372	0.0138	0.0489	18.6366

Table A7. Measure group 1 results on corpora PUA.

NB		WAcc	WErr	WRej	TCR
$\lambda = 1$	NB	0.9570	0.0430	0.0000	19.8776
	Three-way	0.9509	0.0377	0.0114	22.1418
$\lambda = 3$	NB	0.9575	0.0425	0.0000	17.2113
	Three-way	0.9500	0.0382	0.0118	19.0486
$\lambda = 9$	NB	0.9577	0.0423	0.0000	16.0920
	Three-way	0.9495	0.0384	0.0121	17.6724
k-nn					
$\lambda = 1$	k-nn	0.7175	0.2825	0.0000	2.2809
	Three-way	0.6237	0.2123	0.1632	4.0057
$\lambda = 3$	k-nn	0.5877	0.4123	0.0000	0.8063
	Three-way	0.4671	0.3197	0.2132	1.3352

Table A7. (*Continued*)

k-nn		WAcc	WErr	WRej	TCR
$\lambda = 9$	k-nn	0.5098	0.4902	0.0000	0.2745
	Three-way	0.3732	0.3837	0.2432	0.4451
SVM					
$\lambda = 1$	SVM	0.9316	0.0684	0.0000	11.7014
	Three-way	0.8728	0.0272	0.1000	32.5111
$\lambda = 3$	SVM	0.9184	0.0816	0.0000	7.8165
	Three-way	0.8548	0.0338	0.1114	23.8442
$\lambda = 9$	SVM	0.9105	0.0895	0.0000	5.9354
	Three-way	0.8440	0.0377	0.1182	20.6893

Table A8. Measure group 1 results on corpora Enron-Spam.

NB		WAcc	WErr	WRej	TCR
$\lambda = 1$	NB	0.8583	0.1417	0.0000	3.9030
	Three-way	0.8447	0.1293	0.0260	4.5028
$\lambda = 3$	NB	0.8710	0.1290	0.0000	2.6236
	Three-way	0.8589	0.1179	0.0232	3.0856
$\lambda = 9$	NB	0.8620	0.1380	0.0000	2.0322
	Three-way	0.8503	0.1270	0.0227	2.4331
k-nn					
$\lambda = 1$	k-nn	0.7726	0.2274	0.0000	37.9166
	Three-way	0.7598	0.2143	0.0260	44.1145
$\lambda = 3$	k-nn	0.7531	0.2469	0.0000	17.7413
	Three-way	0.7359	0.2316	0.0325	18.9961
$\lambda = 9$	k-nn	0.7137	0.2863	0.0000	7.0077
	Three-way	0.6928	0.2688	0.0385	7.1975
SVM					
$\lambda = 1$	SVM	0.9393	0.0607	0.0000	9.8972
	Three-way	0.8886	0.0272	0.0843	23.0962
$\lambda = 3$	SVM	0.9371	0.0629	0.0000	6.4686
	Three-way	0.8932	0.0279	0.0790	13.7574
$\lambda = 9$	SVM	0.9321	0.0679	0.0000	4.4352
	Three-way	0.8920	0.0300	0.0780	8.5106

Table A9. Four group of cost functions used in the experiment

Group number	λ_{PN}	λ_{BN}	λ_{BP}	λ_{NP}	α	β
Group 1	0.3333	0.1667	0.1667	0.3333	0.5000	0.5000
Group 2	0.0840	0.0672	0.0084	0.8403	0.5294	0.0108
Group 3	0.4785	0.0383	0.0048	0.4785	0.9252	0.0108
Group 4	0.8475	0.0678	0.0085	0.0763	0.9252	0.5000

Table A10. Accuracy and error rate on corpora Ling-Spam-stop.

Cost functions	Classifier	SOLU-THREE		SOLU-TWO		SOLU-TWO-COST	
		Acc	Err	Acc	Err	Acc	Err
Group 1	NB	0.9007	**0.0993**	0.9007	**0.0993**	0.9007	**0.0993**
	C4.5	0.9128	**0.0993**	0.9007	**0.0993**	0.9007	**0.0993**
	k-nn	0.9093	0.0907	0.9099	**0.0901**	0.9099	**0.0901**
Group 2	NB	0.8595	**0.0879**	0.9007	0.0993	0.8336	0.1664
	C4.5	0.8882	**0.0820**	0.9128	0.0872	0.8526	0.1474
	k-nn	0.8193	**0.0114**	0.9099	0.0901	0.9266	0.0733
Group 3	NB	0.8419	**0.0612**	0.9007	0.0993	0.9007	0.0933
	C4.5	0.8879	**0.0820**	0.9128	0.0872	0.9128	0.0872
	k-nn	0.7345	**0.0083**	0.9099	0.0901	0.9099	0.0901
Group 4	NB	0.8830	**0.0716**	0.9007	0.0993	0.7892	0.2108
	C4.5	0.9125	**0.0872**	0.9128	0.0872	0.7986	0.2014
	k-nn	0.7805	**0.0710**	0.9099	0.0901	0.7996	0.2004

Table A11. Accuracy and error rate on corpora PU1.

Cost functions	Classifier	SOLU-THREE		SOLU-TWO		SOLU-TWO-COST	
		Acc	Err	Acc	Err	Acc	Err
Group 1	NB	0.9211	**0.0789**	0.9211	**0.0789**	0.9211	**0.0789**
	C4.5	0.9211	**0.0789**	0.9211	**0.0789**	0.9211	**0.0789**
	k-nn	0.9083	**0.0917**	0.9083	**0.0917**	0.9083	**0.0917**
Group 2	NB	0.9073	**0.0780**	0.9211	0.0789	0.6807	0.3192
	C4.5	0.7963	**0.0679**	0.9211	0.0789	0.8339	0.1661
	k-nn	0.8972	**0.0826**	0.9083	0.0917	0.9110	0.0890
Group 3	NB	0.9073	**0.0661**	0.9211	0.0789	0.9211	0.0789
	C4.5	0.7917	**0.0661**	0.9211	0.0789	0.9211	0.0789
	k-nn	0.8936	**0.0826**	0.9083	0.0917	0.9083	0.0917
Group 4	NB	0.9211	**0.0670**	0.9211	0.0789	0.6789	0.3211
	C4.5	0.9165	**0.0771**	0.9211	0.0789	0.8844	0.1156
	k-nn	0.8954	**0.0872**	0.9083	0.0917	0.9000	0.1000

Table A12. Accuracy and error rate on corpora PU2.

Cost functions	Classifier	SOLU-THREE		SOLU-TWO		SOLU-TWO-COST	
		Acc	*Err*	*Acc*	*Err*	*Acc*	*Err*
Group 1	NB	0.8803	**0.1197**	0.8803	**0.1197**	0.8803	**0.1197**
	C4.5	0.9549	**0.0451**	0.9549	**0.0451**	0.9549	0.0451
	k-nn	0.9493	**0.0507**	0.9493	**0.0507**	0.9493	0.0507
Group 2	NB	0.8789	**0.1197**	0.8803	**0.1197**	0.8183	0.1817
	C4.5	0.9493	**0.0394**	0.9549	0.0451	0.8521	0.1479
	k-nn	0.9423	**0.0479**	0.9493	0.0507	0.9507	0.0493
Group 3	NB	0.8789	**0.1197**	0.8803	**0.1197**	0.8803	**0.1197**
	C4.5	0.9437	**0.0394**	0.9549	0.0451	0.9543	0.0451
	k-nn	0.9324	**0.0479**	0.9493	0.0507	0.9493	0.0507
Group 4	NB	0.8803	0.1197	0.8803	0.1197	0.9605	**0.0394**
	C4.5	0.9493	**0.0451**	0.9549	**0.0451**	0.8901	0.1099
	k-nn	0.9324	**0.0493**	0.9493	0.0507	0.9338	0.0662

Table A13. Accuracy and error rate on corpora PUA.

Cost functions	Classifier	SOLU-THREE		SOLU-TWO		SOLU-TWO-COST	
		Acc	*Err*	*Acc*	*Err*	*Acc*	*Err*
Group 1	NB	0.9570	**0.0430**	0.9570	**0.0430**	0.9570	**0.0430**
	C4.5	0.8754	**0.1246**	0.8754	**0.1246**	0.8754	**0.1246**
	k-nn	0.7640	**0.2360**	0.7640	**0.2360**	0.7640	**0.2360**
Group 2	NB	0.9474	**0.0351**	0.9570	0.0430	0.5947	0.4053
	C4.5	0.3982	**0.0219**	0.8754	0.1247	0.7553	0.2447
	k-nn	0.7421	**0.2263**	0.7640	0.2360	0.7693	0.2307
Group 3	NB	0.9486	**0.0316**	0.9570	0.0430	0.9570	0.0430
	C4.5	0.3947	**0.0211**	0.8754	0.1246	0.8754	0.1246
	k-nn	0.7395	**0.2263**	0.7640	0.2360	0.7640	0.2360
Group 4	NB	0.9526	**0.0395**	0.9570	0.0430	0.6070	0.3930
	C4.5	0.8719	**0.1237**	0.8754	0.1246	0.8342	0.1658
	k-nn	0.7404	**0.2325**	0.7640	0.2360	0.7439	0.2561

Table A14. Misclassification cost on corpora Ling-Spam-stop.

Cost functions	Classifier	SOLU-THREE	SOLU-TWO	SOLU-TWO-COST
Group 1	NB	**9.5667**	**9.5667**	**9.5667**
	C4.5	**8.4000**	**8.4000**	**8.4000**
	k-nn	8.7333	**8.6667**	**8.6667**
Group 2	NB	**2.5899**	21.6975	40.4201
	C4.5	**3.5134**	18.6807	35.7983
	k-nn	**3.6462**	7.0252	16.6807
Group 3	NB	**8.8742**	13.7321	13.7321
	C4.5	**11.4349**	12.0574	12.0574
	k-nn	**3.3627**	12.4402	12.4402
Group 4	NB	15.4881	**4.4568**	50.8389
	C4.5	18.8178	**4.4669**	48.4737
	k-nn	**4.8398**	17.0983	48.3737

Table A15. Misclassification cost on corpora PU1.

Cost functions	Classifier	SOLU-THREE	SOLU-TWO	SOLU-TWO-COST
Group 1	NB	**2.8667**	**2.8667**	**2.8667**
	C4.5	**2.8667**	**2.8667**	**2.8667**
	k-nn	**3.3333**	**3.3333**	**3.3333**
Group 2	NB	**1.0361**	6.8487	28.7899
	C4.5	**1.8**	5.3361	14.3782
	k-nn	**4.2664**	4.6218	4.7479
Group 3	NB	**3.4622**	4.1148	4.1148
	C4.5	**3.5761**	4.1148	4.1148
	k-nn	**4.3823**	4.7847	4.7847
Group 4	NB	5.8119	**1.0415**	29.3525
	C4.5	5.2263	**2.5839**	10.2153
	k-nn	**4.2941**	4.6186	5.7669

Table A16. Misclassification cost on corpora PU2.

Cost functions	Classifier	SOLU-THREE	SOLU-TWO	SOLU-TWO-COST
Group 1	NB	**2.8333**	**2.8333**	**2.8333**
	C4.5	**1.0667**	**1.0667**	**1.0667**
	k-nn	**1.2000**	**1.2000**	**1.2000**
Group 2	NB	**1.6983**	6.1597	10.4622
	C4.5	**0.4924**	2.1597	8.8235
	k-nn	**1.0076**	2.2689	2.2605
Group 3	NB	4.0674	**4.0670**	**4.0670**
	C4.5	**1.3722**	1.5311	1.5311
	k-nn	**1.6770**	1.7225	1.7225
Group 4	NB	6.2008	**1.6508**	2.0644
	C4.5	2.1992	**0.7839**	6.1475
	k-nn	2.2771	**1.0458**	2.0551

Table A17. Misclassification cost on corpora PUA.

Cost functions	Classifier	SOLU-THREE	SOLU-TWO	SOLU-TWO-COST
Group 1	NB	**1.6333**	**1.6333**	**1.6333**
	C4.5	**4.7333**	**4.7333**	**4.7333**
	k-nn	**8.9667**	**8.9667**	**8.9667**
Group 2	NB	**1.5462**	2.3025	37.0084
	C4.5	**1.4538**	3.0840	21.1765
	k-nn	21.5151	**2.9412**	2.9664
Group 3	NB	**1.7828**	2.3445	2.3445
	C4.5	**1.8722**	6.7943	6.7943
	k-nn	**12.4770**	12.8708	12.8708
Group 4	NB	**2.0000**	2.2246	36.8864
	C4.5	**2.9543**	10.1059	15.6317
	k-nn	**2.5932**	22.1025	24.3602

References

1. Androutsopoulos, I., Paliouras, G., Karkaletsis, V., Sakkis, G., Spyropoulos, C.D., Stamatopoulos, P.: Learning to filter spam e-mail: a comparison of a naive bayesian and a memory-based approach. In: 4th European Conference on Principles and Practice of Knowledge Discovery in Databases, pp. 1–13 (2000)
2. Benevenuto, F., Magno, G., Rodrigues, T., Almeida, V.: Detecting spammers on twitter. In: 7th Annual Collaboration, Electronic Messaging, Anti-Abuse and Spam Conference (2010)
3. Carreras, X., Marquez, L.: Boosting trees for anti-spam email filtering. In: European Conference on Recent Advances in NLP (2001)
4. Chow, C.K.: On optimum recognition error and reject tradeoff. IEEE Trans. Inf. Theory **16**(1), 41–46 (1970)
5. DeBarr, D., Wechsler, H.: Spam detection using clustering, random forests, and active learning. In: 6th Conference on Email and Anti-Spam (2009)
6. Domingos, P., Pazzani, M.: Beyond independence: conditions for the optimality of the simple Bayesian classifier. In: 13th International Conference on Machine Learning, pp. 105–112 (1996)
7. Drucker, H., Wu, D.H., Vapnik, V.N.: Support vector machines for spam categorization. IEEE Trans. Neural Netw. **10**(5), 1048–1054 (1999)
8. Elkan, C.: The foundations of cost-sensitive learning. In: 17th International Joint Conference on Artificial Intelligence, pp. 973–978 (2001)
9. Fiumara, G., Marchi, M., Pagano, R., Provetti, A., Spada, N.: A rule-based system for end-user e-mail annotations. In: 8th Annual Collaboration, Electronic Messaging, Anti-Abuse and Spam Conference, pp. 102–108 (2011)
10. Halder, S., Wei, C., Sprague, A.: Overview of existing methods of spam mining and potential usefulness of sender's name as parameter for fuzzy string matching. ACEEE Int. J. Inf. Techonol. **1**(1), 25–28 (2011)
11. Hall, M., Frank, E., Holmes, G., Pfahringer, B., Reutemann, P., Witten, I.H.: The WEKA data mining software: an update. SIGKDD Explor. **11**(1), 130–133 (2009)
12. Khonji, M., Jones, A., Iraqi, Y.: A study of feature subset evaluators and feature subset searching methods for phishing classification. In: 8th Annual Collaboration, Electronic Messaging, Anti-Abuse and Spam Coference, pp. 135–144 (2011)
13. Khorsi, A.: An overview of content-based spam filtering techniques. Informatica **31**, 269–277 (2007)
14. Metsis, V., Androutsopoulos, I., Paliouras, G.: Spam filtering with naive bayes-which naive bayes? In: 3rd Conference on Email and Anti-Spam (2006)
15. Mitchell, T.M.: Machine Learning. McGraw-Hill, New York (1997)
16. Pauker, S.G., Kassirer, J.P.: The threshold approach to clinical decision making. N. Engl. J. Med. **302**, 1109–1117 (1980)
17. Pawlak, Z.: Rough sets. Int. J. Comput. Inf. Sci. **11**, 341–356 (1982)
18. Perez-Diaz, N., Ruano-Ordas, D., Mendez, J.R., Galvez, J.F., Fdez-Riverola, F.: Rough sets for spam filtering: selecting appropriate decision rules for boundary e-mail classification. Appl. Soft Comput. **12**(11), 3671–3682 (2012)
19. Sahami, M., Dumais, S., Heckerman, D., Horvitz, E.: A bayesian approach to filtering junk e-mail. In: Learning for Text Categorization-Papers from the AAAI Workshop, pp. 55–62 (1996)
20. Schneider, K.M.: A comparison of event models for Naive Bayes anti-spam e-mail filtering. In: 10th Conference of the European Chapter of the Association for Computational Linguistics, pp. 307–314 (2003)

21. Wei, C., Sprague, A., Warner, G., Skjellum, A.: Identifying new spam domains by hosting IPs: improving domain blacklisting. In: 7th Annual Collaboration, Electronic Messaging, Anti-Abuse and Spam Coference (2010)
22. Yao, Y.Y., Wong, S.K.M., Lingras, P.: A decision-theoretic rough set model. Methodol. Intell. Syst. **5**, 17–24 (1992)
23. Yao, Y.Y.: Three-way decisions with probabilistic rough sets. Inf. Sci. **180**, 341–353 (2010)
24. Zhao, W., Zhu, Y.: An email classification scheme based on decision-theoretic rough set theory and analysis of email security. In: 2005 IEEE Region 10 TENCON, pp. 1–C6 (2005)
25. Zhou, B., Yao, Y.Y., Luo, J.G.: A three-way decision approach to email spam filtering. In: 23th Canadian Conference on Artificial Intelligence, pp. 28–39 (2010)

A Three-Way Decisions Approach
to Density-Based Overlapping Clustering

Hong Yu$^{(\boxtimes)}$, Ying Wang, and Peng Jiao

Chongqing Key Laboratory of Computational Intelligence,
Chongqing University of Posts and Telecommunications, Chongqing 400065, China
yuhong@cqupt.edu.cn

Abstract. Most of clustering methods assume that each object must be
assigned to exactly one cluster, however, overlapping clustering is more
appropriate than crisp clustering in a variety of important applications
such as the network structure analysis and biological information. This
paper provides a three-way decisions approach for overlapping clustering
based on the decision-theoretic rough set model, where each cluster is
described by an interval set which is defined by a pair of sets called the
lower and upper bounds, and the overlapping objects usually are distributed in the region between the lower and upper regions. Besides, a density-
based clustering algorithm is proposed using the approach considering the
advantages of the density-based clustering algorithms in finding the arbitrary shape clusters. The results of comparison experiments show that the
three-way decisions approach is not only effective to overlapping clustering but also good at discovering the arbitrary shape clusters.

Keywords: Overlapping clustering · Three-way decisions · Decision-
theoretic rough set theory · Density-based clustering · Data mining

1   Introduction

In recent years, clustering has been widely used as a powerful tool to reveal under-
lying patterns in many areas such as data mining, web mining, geographical data
processing, medicine and so on. Clustering is the task of grouping a set of objects
in such a way that objects in the same group are more similar to each other than to
those in other groups. Most of clustering methods assume that each object must
be assigned to exactly one cluster. However, in a variety of important applications
such as network structure analysis, wireless sensor networks and biological information, overlapping clustering is more appropriate [7].

Many researchers have proposed some overlapping clustering methods for
different application backgrounds. For example, Obadi et al. [13] proposed an
overlapping clustering method for DBLP datasets based on rough sets theory,
Takaki [19] proposed a method of overlapping clustering for network structure
analysis, Aydin et al. [2] proposed an overlapping clusters algorithm used in
the mobile Ad hoc networks, and Lingras et al. [9] compared crisp and fuzzy
clustering in the mobile phone call dataset.

© Springer-Verlag Berlin Heidelberg 2014
J.F. Peters et al. (Eds.): Transactions on Rough Sets XVIII, LNCS 8449, pp. 92–109, 2014.
DOI: 10.1007/978-3-662-44680-5_6

There are also some achievements by combining uncertain approaches such as fuzzy sets theory and rough sets theory. For example, Lingras and Yao [10] provided comparison of the time complexity of the two rough clustering algorithms based on Genetic Algorithms and K-means algorithm. Asharaf and Murty [1] proposed a rough set theory-based adaptive rough fuzzy leader clustering algorithm, where the upper and lower thresholds decided by the experiential experts. Wu and Zhou [22] proposed a possibilistic fuzzy algorithm based on c-means clustering. Yu and Luo [28] proposed a leader clustering algorithm based on possibility theory and fuzzy theory.

Density-based clustering is a typical successful clustering algorithm [17,18]. The density-based algorithm is a natural and attractive for data streams, because it can not only find arbitrarily shaped cluster and handle noises, but also it is an one-scan algorithm which needs to scan the raw data only one time. DBSCAN, proposed by Ester et al. in 1996 [6], was the first clustering algorithm to find a number of clusters starting from the estimated density distribution of corresponding nodes. The algorithm set two parameters in advance, the radius and minimum number of included points. However, to define the parameters are difficult, and which also affects the clustering result. KIDBSCAN was another density-based clustering method presented by Tsai and Liu in 2006 [20], in which K-means is used to find the high-density center points and then IDBSCAN is used to expand clusters from these high-density center points. Duan et al. [5] proposed a clustering algorithm LDBSCAN which relying on a local-density-based notion of cluster. Yousri et al. proposed a novel possibilistic density based clustering approach [27] to identify the degrees of typicality of patterns to clusters of arbitrary shapes and densities. Ren et al. [16] proposed an effective merging approach by using the local sub-cluster density information. Furthermore, in most of density-based clustering algorithms, the noise data are simply deleted or does not dealt, which is not quite up to the factual situations.

To observe these clustering algorithms, we can find that almost of them are based on two-way decisions, that is, the decision to determine one object whether belong to a cluster is accepted or rejected. However, in many practical applications, due to the inaccuracy of the information or not integrity, the decision can not be done as acceptance or rejection. In other words, it is difficult to accept or not to accept, and it is difficult to reject or not reject. In fact, we often defer decisions in this case, where the three-way decisions method is used unconsciously.

The rough sets theory [14] approximates a concept by three regions, namely, the positive, boundary and negative regions, which immediately leads to the notion of three-way decisions clustering approach. Three-way decisions constructed from the three regions are associated with different actions and decisions. In fact, the three-way decisions approach has been achieved in some areas as the email spam filtering [29], three-way investment decisions [11], and so on [8,24].

When talking of rough sets, the classical Pawlak lower and upper approximations are defined based on qualitative set-inclusion and non-empty overlapping relations, respectively. Consequently, the theory suffers from an intolerance of errors, which greatly restricts its real-world applications. To remedy this limitation, Yao and associates [26] proposed the decision-theoretic rough sets (DTRS) model in early 1990s' by introducing the Bayesian decision theory into rough sets. Corresponding to the three regions, Yao [24] introduced and studied the notion of three-way decisions, consisting of the positive, boundary and negative rules. The positive rules generated by the positive region make decision of acceptance, the negative rules generated by the negative region make decisions of rejection, and the boundary rules generated by the boundary region make deferred or non-committed decisions.

Therefore, this paper proposes a new three-way decisions clustering approach based on the decision-theoretic rough set model to combat the overlapping clustering. Yao et al. [25] have represented each cluster by an interval set instead of a single set as the representation of a cluster. Chen and Miao [4] studied the clustering method represented as interval sets, wherein the rough k-means clustering method is combined. Inspired by the representation, the cluster in this paper is also represented by an interval set, which is defined by a pair of sets called the lower and upper bounds. Objects in the lower bound are typical elements of the cluster and objects between the upper and lower bounds are fringe elements of the cluster. Furthermore, the solutions to obtain the lower and upper bounds are formulated based on the three-way decisions in this paper. Then, a density-based clustering algorithm is proposed, and we demonstrate the validity of the algorithm through experiments.

The rest of the paper is organized as follows. Some basic concepts about the three-way decisions theory and the decision-theoretic rough set model are introduced in the next session. Section 3 extends the decision-theoretic rough set model to clustering firstly, then re-formulates the representation of clustering by using three-way decisions and wherein an interval set is used to depict a cluster in order to formulate the overlapping region. A density-based overlapping clustering algorithm using three-way decisions is proposed in Sect. 4. Then, a synthetic data set, the Chameleon data sets [3] and the dolphin social network data [30] are used in experiments in Sect. 5. Some conclusions will be given in the final section.

2   Preliminaries

The concept of three-way decisions plays an important role in many real world decision-making problems. One usually makes a decision based on available information and evidence. When the evidence is insufficient or weak, it might be impossible to make either a positive or a negative decision. One therefore chooses an alternative decision that is neither yes nor no. A few examples are given as illustrations. In the editorial peer review process of many journals, an editor typically makes a three-way decision about a manuscript, namely, acceptance, rejection, or further revision, based on comments from reviewers; the final decision of the third category is subject to comments from reviews in another round of review [21].

The three-way decisions rules come closer to the philosophy of the rough set theory [23], namely, representing a concept using three regions instead of two. This three-way decisions scheme has not been considered explicitly in other theories of machine learning and rule induction, although it has been studied in other fields. By considering three-way decision rules, one may appreciate the true and unique contributions of rough set theory to machine learning and rule induction.

Decision-theoretic rough set model was proposed by Yao and associates in 1990s', which gives the three-way decisions a representation. Let's review some basis concepts, notations and results of DTRS in the following.

An information table is the objective of study, which can be seen as a data table where columns are labeled by attributes, rows are labeled by objects and each row represents some information about the corresponding object. That is, an information table is a quadruple: $S = (U, A, \{V_a | a \in A\}, \{I_a | a \in A\})$, where U is a finite nonempty set of objects, A is a finite nonempty set of attributes, V_a is a nonempty set of values of $a \in A$, and $I_a : U \longrightarrow V_a$ is an information function that maps an object in U to exactly one value in V_a.

Let $\Omega = \{X, X^c\}$ denotes the set of states indicating that an object is in X and not in X, respectively. Let $Action = \{a_P, a_N, a_B\}$ be the set of actions, where a_P, a_N, and a_B represent the three actions in classifying an object, deciding $POS(X)$, deciding $NEG(X)$ and deciding $BND(X)$, respectively. Let $i = P, N, B$, and $\lambda_{iP}(a_i|A)$ and $\lambda_{iN}(a_i|A^c)$ denote the loss (cost) for taking the action a_i when the state is X, X^c, respectively. For an object with description $[x]$, suppose an action a_i is taken, the expected loss $R(a_i|[x])$ associated with taking the individual actions can be expressed as:

$$
\begin{aligned}
R(a_P|[x]) &= \lambda_{PP}P(X|[x]) + \lambda_{PN}P(X^c|[x]), \\
R(a_N|[x]) &= \lambda_{NP}P(X|[x]) + \lambda_{NN}P(X^c|[x]), \\
R(a_B|[x]) &= \lambda_{BP}P(X|[x]) + \lambda_{BN}P(X^c|[x]).
\end{aligned} \tag{1}
$$

Where the probabilities $P(X|[x])$ and $P(X^c|[x])$ are the probabilities that an object in the equivalence class $[x]$ belongs to X and X^c, respectively.

The Bayesian decision procedure leads to the following minimum-risk decision:

$$
\begin{aligned}
&(P) If R(a_P|[x]) \leq R(a_N|[x]) \ and \ R(a_P|[x]) \leq R(a_B|[x]), \\
&decide \ x \in POS(X); \\
&(B) If \ R(a_B|[x]) < R(a_P|[x]) \ and \ R(a_B|[x]) < R(a_N|[x]), \\
&decide \ x \in BND(X); \\
&(N) If \ R(a_N|[x]) \leq R(a_P|[x]) \ and \ R(a_N|[x]) \leq R(a_B|[x]), \\
&decide \ x \in NEG(X).
\end{aligned} \tag{2}
$$

3 Formulation of Clustering

In this section, the representation of clustering will be re-formulated by using three-way decisions based on the extended decision-theoretic rough set model, wherein an interval set is used to depict a cluster in order to formulate the overlapping region.

3.1   Extend DTRS for Clustering

A main task of cluster analysis is to group objects in a universe so that objects in the same cluster are more similar to each other and objects in different clusters are dissimilar.

To define our framework, we will assume $\mathbf{C} = \{C_1, \cdots, C_k, \cdots, C_K\}$, where $C_k \subseteq U$, is a family of clusters of a universe $U = \{x_1, \cdots, x_n\}$.

In order to interpret clustering, let's extend the DTRS model firstly. The set of states is given by $\Omega = \{C, \neg C\}$, the two complement states indicate that an object is in a cluster C and not in a cluster C, respectively. The set of action is given by $A = \{a_P, a_B, a_N\}$, where a_P, a_B and a_N represent the three actions in classifying an object, a_P represents that we will take the description of an object x into the domain of the cluster C; a_B represents that we will take the description of an object x into the boundary domain of the cluster C; a_N represents that we will take the description of an object x into the negative domain of the C.

Let $\lambda_{PP}, \lambda_{BP}, \lambda_{NP}, \lambda_{PN}, \lambda_{BN}, \lambda_{NN}$ denote the loss (cost) for taking the action a_P, a_B and a_N when the state is C, $\neg C$, respectively. For an object x with description $[x]$, suppose an action a_i is taken. According to Subsect. 2.1, the expected loss associated with taking the actions can be expressed as:

$$Risk(a_P|[x]) = \lambda_{PP} Pr(C|[x]) + \lambda_{PN} Pr(\neg C|[x]);$$
$$Risk(a_B|[x]) = \lambda_{BP} Pr(C|[x]) + \lambda_{BN} Pr(\neg C|[x]); \qquad (3)$$
$$Risk(a_N|[x]) = \lambda_{NP} Pr(C|[x]) + \lambda_{NN} Pr(\neg C|[x]).$$

Where $Pr(C|[x])$ represents the probability that an object x in the description $[x]$ belongs to the cluster C, and $Pr(C|[x]) + Pr(\neg C|[x]) = 1$. The Bayesian decision procedure leads to the following minimum-risk decision:

$$(P) If\ Risk(a_P|[x]) \leq Risk(a_N|[x])\ and\ Risk(a_P|[x]) \leq Risk(a_B|[x]),$$
decide $POS(C)$;
$$(B) If\ Risk(a_B|[x]) < Risk(a_P|[x])\ and\ Risk(a_B|[x]) < Risk(a_N|[x]),$$
decide $BND(C)$; $\qquad (4)$
$$(N) If\ Risk(a_N|[x]) \leq Risk(a_P|[x])\ and\ Risk(a_N|[x]) \leq Risk(a_B|[x]),$$
decide $NEG(C)$.

Consider a special kind of loss functions with $\lambda_{PP} \leq \lambda_{BP} < \lambda_{NP}$ and $\lambda_{NN} \leq \lambda_{BN} < \lambda_{PN}$. That is, the loss of classifying an object x belonging to C into the positive region $POS(C)$ is less than or equal to the loss of classifying x into the boundary region $BND(C)$, and both of these losses are strictly less than the loss of classifying x into the negative region $NEG(C)$. The reverse order of losses is used for classifying an object x that does not belong to C, namely the object x is a negative instance of C. For this type of loss function, the above minimum-risk decision rules can be written as:

$$(P) If\ Pr(C|[x]) \geq \alpha\ and\ Pr(C|[x]) \geq \gamma, decide\ POS(C);$$
$$(B) If\ Pr(C|[x]) < \alpha\ and\ Pr(C|[x]) > \beta, decide\ BND(C); \qquad (5)$$
$$(N) If\ Pr(C|[x]) \leq \beta\ and\ Pr(C|[x]) \leq \gamma, decide\ NEG(C).$$

Where:

$$\alpha = \frac{(\lambda_{PN}-\lambda_{BN})}{(\lambda_{PN}-\lambda_{BN})+(\lambda_{BP}-\lambda_{PP})} = (1+\frac{(\lambda_{BP}-\lambda_{PP})}{(\lambda_{PN}-\lambda_{BN})})^{-1}$$
$$\gamma = \frac{(\lambda_{PN}-\lambda_{NN})}{(\lambda_{PN}-\lambda_{NN})+(\lambda_{NP}-\lambda_{PP})} = (1+\frac{(\lambda_{NP}-\lambda_{PP})}{(\lambda_{PN}-\lambda_{NN})})^{-1} \qquad (6)$$
$$\beta = \frac{(\lambda_{BN}-\lambda_{NN})}{(\lambda_{BN}-\lambda_{NN})+(\lambda_{NP}-\lambda_{BP})} = (1+\frac{(\lambda_{NP}-\lambda_{BP})}{(\lambda_{BN}-\lambda_{NN})})^{-1}$$

In this paper, we consider that the cluster have the boundary, so we just discuss the relationship between thresholds α and β as $\alpha > \beta$. According to Eq. (4), it follows that $\alpha > \gamma > \beta$. After tie-breaking, the following simplified rules (P)–(N) are obtained:

$$(P) \; If \; Pr(C|[x]) \geq \alpha, decide \; POS(C);$$
$$(B) \; If \; \beta < Pr(C|[x]) < \alpha, decide \; BND(C); \qquad (7)$$
$$(N) \; If \; Pr(C|[x]) \leq \beta, decide \; NEG(C).$$

Obviously, rules (P)–(N) give a three-way decision method for clustering. That is, an object belongs to a cluster definitely if it is in $POS(C)$ based on the available information; an object may be a fringe member if it is in $BND(C)$, we can decide whether it is in a cluster through further information. Clustering algorithms can be devised according to the rules (P)–(N).

On the other hand, according to the rough set theory [15] and the rules (P)–(N), for a subset $C \subseteq U$, we can define its lower and upper approximations as follows.

$$\underline{apr}(C) = POS(C) = \{x|Pr(C|[x]) \geq \alpha\};$$
$$\overline{apr}(C) = POS(C) \cup BND(C) = \{x|Pr(C|[x]) > \beta\}. \qquad (8)$$

3.2   Re-formulation of Clustering Using Interval Sets

Yao et al. [25] had formulated the clustering using the form of interval sets. It is natural that the region between the lower and upper bound of an interval set means the overlapping region.

Assume $\mathbf{C} = \{C_1, \cdots, C_k, \cdots, C_K\}$ is a family of clusters of a universe $U = \{x_1, \cdots, x_n\}$. Formally, we can define a clustering by the properties:

$$(i) \; C_k \neq \emptyset, 0 \leq k \leq K; \qquad (ii) \; \bigcup_{C_k \in \mathbf{C}} C_k = U.$$

Property (i) requires that each cluster cannot be empty. Property (ii) states that every $x \in U$ belongs to at least one cluster. Furthermore, if $C_i \cap C_j = \emptyset, i \neq j$, it is a crisp clustering, otherwise it is an overlapping clustering.

As we have discussed, we may use an interval set to represent the cluster in $\mathbf{C}$, namely, C_k is represented by an interval set $[C_k^l, C_k^u]$. Combine the conclusion in the above subsection, we can represent the lower and upper bound of the interval set as the lower and upper approximates, that is, C_k is represented by an interval set $[\underline{apr}(C_k), \overline{apr}(C_k)]$.

Any set in the family $[\underline{apr}(C_k), \overline{apr}(C_k)] = \{X|\underline{apr}(C_k) \subseteq X \subseteq \overline{apr}(C_k)\}$ may be the actual cluster C_k. The objects in $\underline{apr}(C_k)$ may represent typical objects of the cluster C_k, objects in $\overline{apr}(C_k) - \underline{apr}(C_k)$ may represent fringe objects, and objects in $U - \overline{apr}(C_k)$ may represent the negative objects. With respect to the family of clusters $\mathbf{C} = \{C_1, \cdots, C_k, \cdots, C_K\}$, we have the following family of interval set clusters:

$$\mathbf{C} = \{[\underline{apr}(C_1), \overline{apr}(C_1)], \ldots, [\underline{apr}(C_k), \overline{apr}(C_k)], \ldots, [\underline{apr}(C_K), \overline{apr}(C_K)]\}.$$

Corresponding to Property (i) and (ii), we adopt the following properties for a clustering in the form of interval set:

$$(i) \; \underline{apr}(C_k) \neq \emptyset, 0 \leq k \leq K; \qquad (ii) \; \bigcup \overline{apr}(C_k) = U.$$

Property (i) requires that the lower approximate must not be empty. It implies that the upper approximate is not empty. It is reasonable to assume that each cluster must contain at least one typical object and hence its lower bound is not empty. In order to make sure that a clustering is physically meaningful, Property (ii) states that any object of U belongs to the upper approximate of a cluster, which ensures that every object is properly clustered.

According to Eq. (6), the family of clusters $\mathbf{C}$ give a three-way decision clustering. Namely, objects in $\underline{apr}(C_k)$ are decided definitely to belong to the cluster C_k, objects in $U - \overline{apr}(C_k)$ can be decided not to belong to the cluster C_k. Set $BND(C_k) = \overline{apr}(C_k) - \underline{apr}(C_k)$. Objects in the region $BND(C_k)$ may be belong to the cluster or not.

There exists $k \neq t$, it is possible that $\underline{apr}(C_k) \cap \underline{apr}(C_t) \neq \emptyset$, or $BND(C_k) \cap BND(C_t) \neq \emptyset$. In other words, it is possible that an object belongs to more than one cluster.

4   Clustering Algorithm Using Three-way Decisions

Density-based clustering analysis is one kind of clustering analysis methods that can discover clusters with arbitrary shape and is insensitive to noise data. Therefore, a density-based overlapping clustering algorithm using three-way decision (shorted by DOCA-TWD algorithm) is proposed in this paper, which is a two-step clustering algorithm. That is, the framework of density-based clustering is used firstly to obtain the family of representing regions and the family of noise data sets, which is called the initial clustering processing. Then, according to the three-way decision rules (P)–(N) in Subsect. 3.1, a final clustering result is obtained to combat the overlapping clustering, which is called the three-way decisions clustering processing. In addition, the approach will decide the outliers to a boundary of the most neighboring cluster or to be noise data by judging the merging possibility, rather than simply deleting or merging them together.

4.1   Basic Concepts

In this paper, the framework of density-based clustering algorithm is used in the first processing. However, in the density-based clustering algorithm, the generation of a reference point is sensitive to choosing the initial point, and the different choices will lead to different clustering results. In order to reduce the influence of the choosing initial points to the reference points, the initial points are calculated based on the grid computing method. Thus, the subspace is a basic concept used here.

Let the $GRID$ is the original data set space, the universe is divided into some subspaces such as $GRID_1, GRID_2, \cdots, GRID_T$ according to the distance threshold value Rth. A subspace can be seen as a partition of the universe, and the intersection of two arbitrary subspace is empty, that is $GRID_i \cap GRID_j = \emptyset$; the union of all subspace is the whole space, that is $GRID_1 \cup GRID_2 \cup \cdots \cup GRID_i \cup \cdots \cup GRID_T = U$. The definition of reference points and representing region are defined as follows [12].

Definition 1. *Reference points:* *For any subspace $GRID_t$, and the density threshold value is Mth in the space, if the number of points in the $GRID_t$ is greater than the Mth, then the centroid of the $GRID_t$ is a reference point p of the subspace.*

The reference points are fictional points, not the points in the data set. Threshold value Mth represents a reference number. When the number of points in $GRID_t$ is greater than Mth, denoted by $|GRID_t| > mth$, the centroid of the $GRID$ is a reference point p, otherwise the $GRID$ is a noise data set.

Definition 2. *Representing region:* *Every reference point p is regarded as the representative of a circular region, where the point is the center of the region and the radius is Rth. The region is called the representing region of the reference point p.*

All points (objects) in the representing region of a reference point p are taken as an equivalence class. In order to cluster objects (points) in the space, we need to give the method to calculate the probability, which is described in Eq. (9).

Definition 3. *Probability:* *Set $[x]$ is a description of an object x, the $Pr(C|[x])$ is:*

$$Pr(C|[x]) = \frac{|C \cap [x]|}{|[x]|}. \tag{9}$$

General speaking, the equivalence class $[x]$ of an object x can be used as a description of the object. That is, Eq. (9) gives a computing method for the probability, then we can devise an algorithm based on three-way decisions. In other words, the different algorithms can be developed based on the different approaches of computing probability.

4.2 Initial Clustering Processing

Let's consider an information table $S = \{U, A\}$, where $U = \{x_1, \cdots, x_i, \cdots, x_n\}$, $A = \{a_1, \cdots, a_m, \cdots, a_M\}$. Now we introduce some signs used below firstly.

$\mathbf{GRID} = \{GRID_1, \cdots, GRID_t, \cdots, GRID_T\}$, $P = \{P_1, \cdots, P_t, \cdots, P_T\}$, $\mathbf{RF} = \{RF_1, \cdots, RF_t, \cdots, RF_T\}$ and $\mathbf{NS} = \{NS_1, \cdots, NS_t, \cdots, NS_N\}$ means the subspace of the universe, the set of reference points, the family of representing regions, and the family of noise data sets, respectively; and max_m means the maximal value in the value domain of the corresponding attribute a_m; and a_{im} is the value of the object x_i in the attribute a_m. Rth and Mth are the thresholds of the radius and the reference number, respectively.

Algorithm 1. The Initial Clustering Processing

Input: $S = \{U, A\}$, Rth, Mth.
Output: the family of representing regions: $\mathbf{RF} = \{RF_1, \ldots, RF_t, \ldots, RF_T\}$;
The family of noise data sets: $\mathbf{NS} = \{NS_1, \cdots, NS_t, \cdots, NS_N\}$.
begin

   Step1: Initial: $\mathbf{RF} = \emptyset$, $Pr(C_k|RF_t) = 0$, $\mathbf{GRID} = \emptyset$, $\mathbf{NS} = \emptyset$,
   $max_m = 0$, $P = \emptyset$.
   Step2: Find all subspaces:
   for every a_m do {
      for every x_i do { If $(a_{im} > max_m)$ then $max_m = a_{im}$; }
      $max_m = max_m/Rth$; }
   for every x_i do {
      for every a_m do {
         $temp = temp + a_{im}/Rth \times max_1 \times \ldots \times max_{m-1}$;}
         $GRID_{temp} = GRID_{temp} \cup \{x_i\}$;}
   Step3. Find the optimal reference point sets:
   for every $GRID_t$ do {
      If $|GRID_t| > Mth$ then $P_t = centerpoint(GRID_t)$ else $NS_t = GRID_t$;}
   Step4. Find the optimal Representing region:
   for every P_t do {
      for every x_i do {
         If $|P_t - x_i| < Rth$ then $RF_t = RF_t \cup \{x_i\}$; } }
end

The basic idea of the initial clustering processing is described in Algorithm 1. That is, according to the threshold value Rth, the original data set is divided into many subspaces $GRID_t$ firstly. Then, if the number of points in the subspace $GRID_t$ is greater than Mth, set the center point of $GRID_t$ as the optimal reference point P_t; otherwise, set the $GRID_t$ as a noise set NS_t. Finally, according to Rth and the reference point P_t, the corresponding representative region is built.

4.3   Three-Way Decisions Clustering Processing

From Algorithm 1, we obtain the $\mathbf{RF} = \{RF_1, \cdots, RF_t, \cdots, RF_T\}$ and $\mathbf{NS} = \{NS_1, \cdots, NS_t, \cdots, NS_N\}$; they means the family of representing regions and the family of noise data sets, respectively. Then, the three-way decisions rules are used to expand the representative region and merge the small granularity areas, and the basic idea of the processing of three-way decision is described in Algorithm 2.

Algorithm 2. Three-way Decisions Clustering Algorithm (TDCA)

Input: RF; NS; α, β.
Output: The result of clustering:
$C = \{[\underline{apr}(C_1), \overline{apr}(C_1)], \ldots, [\underline{apr}(C_k), \overline{apr}(C_k)], \ldots, [\underline{apr}(C_K), \overline{apr}(C_K)]\}.$
begin

  Step 1. Clustering the reference points according to three-way rules (P)–(N):
  for every $\underline{apr}(C_k)$ do{
  　$\underline{apr}(C_k) = RF_k;$　$\overline{apr}(C_k) = RF_k;$
  　for every RF_t do {
  　calculate the $Pr(C_k|RF_t)$ according to Eq. (10);
  　if $Pr(C_k|RF_t) \geq \alpha$ then {
  　　$\underline{apr}(C_k) = \underline{apr}(C_k) \cup RF_t;$ $\overline{apr}(C_k) = \overline{apr}(C_k) \cup RF_t;$}
  　if $\beta < Pr(C_k|RF_t) < \alpha$ then　$\overline{apr}(C_k) = \overline{apr}(C_k) \cup RF_t;$ }}
  Step 2. Merging clusters according to three-way rules (P)–(N):
  for every $\underline{apr}(C_i)$ do {
  　for every $\underline{apr}(C_j)$ do {
  　calculate the $Pr(C_i|C_j)$ according to Eq. (11);
  　if $Pr(C_i|C_j) \geq \alpha$ then {
  　　$\underline{apr}(C_i) = \underline{apr}(C_i) \cup \underline{apr}(C_j);$ $\overline{apr}(C_i) = \overline{apr}(C_i) \cup \overline{apr}(C_j);$
  　　$\mathbf{C} = \mathbf{C} - C_j;$}
  　if $\beta \leq Pr(C_i|C_j) < \alpha$ then $\overline{apr}(C_i) = \overline{apr}(C_i) \cup \overline{apr}(C_j);$ }}
  Step 3. Processing the noise points:
  for every NS_t do {
  　for every $\underline{apr}(C_k)$ do {
  　calculate the $Pr(C_k|NS_t)$ according to Eq. (12);
  　if $Pr(C_k|NS_t) \geq \beta$ then {
  　　$\overline{apr}(C_k) = \overline{apr}(C_k) \cup NS_t;$ **NS = NS** $- NS_t;$} } }
  for every NS_t do {
  　$\underline{apr}(C_{K+1}) = \underline{apr}(C_{K+1}) \cup NS_t;$}
end

According to Eq. (9), we extend to define the probability of a cluster to a representative region as follows.

$$Pr(C_k|RF_t) = |\underline{apr}(C_k) \cap RF_t|/|RF_t| \tag{10}$$

Similarly, the merging probability between two lower approximation of clusters is defined as following.

$$Pr(C_i|C_j) = |\underline{apr}(C_i) \cap \underline{apr}(C_j)|/|\underline{apr}(C_i)| \tag{11}$$

And the combining probability of a noise data set to a lower approximation of cluster is defined as following.

$$Pr(C_k|NS_t) = |\underline{apr}(C_k) \cap NS_t|/|NS_t| \tag{12}$$

In the three-way decisions processing, every representative area in **RF** is set to be the upper and lower approximation of a cluster firstly. Then, according to the probability of a cluster to a representative region, the three-way decision rules are used to decide whether the representative region adds to the lower approximation or the boundary of the cluster. After that, according to the merging probability between the two lower approximation of clusters, the three-way decision rules are used to merge the clusters or add the lower approximation of the cluster to the boundary of another one. Finally, the algorithm calculates the combining probability of a noise data set to every lower approximation of cluster, if the probability is greater than the β, the noise data set is added to the boundary of the cluster; then the left noise sets form a new cluster.

5   Experimental Results

In this section, a synthetic data set is used to demonstrate the DOCA-TWD algorithm proposed in this paper firstly, which shows that the method is effective in processing the overlapping clustering and in finding the special points. The special points usually are located in the overlapping boundary of clusters. In some fields such as social networks, it is more significant that to find the special points than to delete them, because these points are often key nodes in the network structure. Then, we apply the method on Chameleon data sets [3] to evaluate its ability in clustering arbitrary shape, different thresholds such as the α and β are also tested and discussed. The algorithms are performed in C++.

5.1   Overlapping Clustering Analysis

The synthetic data set is tested to illustrate the ideas presented in the previous section. The two dimensions data set is depicted in Fig. 1, MD1 have 259 points.

Firstly, set the three thresholds $Rth = 1, \alpha = 0.68, \beta = 0.35$. According to Algorithm 1, the universe is divided into subspaces and the reference points are decided, which are depicted in Fig. 1a and 1b respectively. In the figures, the diamond points represent objects, the circle points represent reference points.

Then, according to Algorithm 2, we obtain the final clustering result in Fig. 1c, and there are five clusters. The region indicated by solid lines is the lower bound (positive) of the corresponding cluster, the region indicated by the dotted line means the upper bound of the corresponding cluster, the region between these is the boundary region.

To observe Fig. 1c, we find that the C_1 and C_4 have no boundary regions, the $POS(C_2)$ overlaps the $POS(C_5)$, the $POS(C_2)$ overlaps the $BND(C_3)$, the $BND(C_2)$ overlaps the $BND(C_3)$, and so on. Let's observe the objects x_1, x_2, x_3, x_4, x_5 in Fig. 1c. Those five points construct the positive of C_5, namely, $POS(C_5)$. Objects x_3, x_4, x_5 belong to the $BND(C_2)$, and x_1, x_2, x_3 belong to the $BND(C_3)$.

The result is interesting, which reminds us to think that we can find some special points through the method proposed in this paper. Here we use special points to denote these points located in the overlapping regions. In some

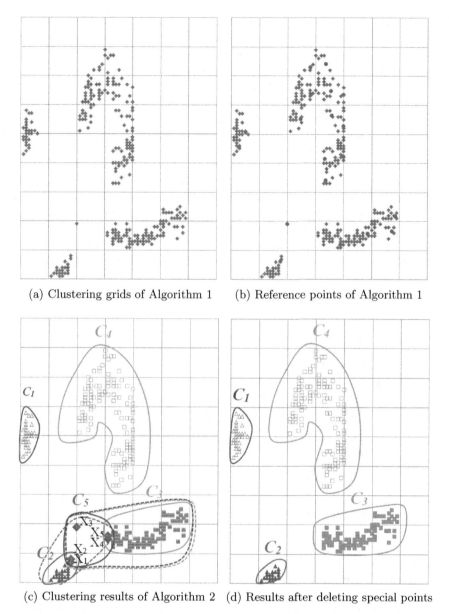

(a) Clustering grids of Algorithm 1   (b) Reference points of Algorithm 1

(c) Clustering results of Algorithm 2   (d) Results after deleting special points

Fig. 1. Clustering results on a synthetic data set MD1

existing clustering methods, these points are often considered the noises or the outliers. But considering some application fields such as social networks, these points often are very important to find the community structure. In order to further illustration, we test again under the same thresholds after deleting the five special points, and the clustering result is shown in Fig. 1d. We can see that the result is different from the result in Fig. 1c. That is, the special points cause the overlapping in this example. In other words, our method can find out the special objects through analyzing the overlapping between regions.

Furthermore, the data set dolphins social networks [30] is used also to verify the ability of processing overlapping. The data set is constructed by Lusseau and his collaborators after seven years of observations (This population include 2 groups which consist of 62 dolphins). Figure 2 shows the networks, one node represents one dolphin, the edge between two nodes signifies that those two dolphins contact with each other frequently. The left shadow points show the positive region of the first group, which has 33 nodes; and the right shadow points show the positive region of the second group, which had 16 nodes; nodes 7, 30, 19 are the overlapping nodes between the positive regions of the two groups, and the node 39 is the overlapping node between the boundaries. Obviously, nodes 7, 30, 19 have more important influence than the node 39. The result just reveal the relationships among dolphins, which is very helpful to discover the evolution of the dolphin community.

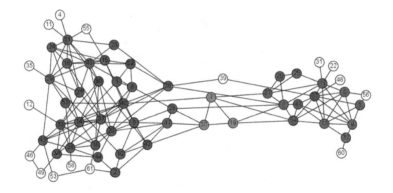

Fig. 2. Clustering results on dolphins social networks

5.2   Arbitrary Shape Clustering Analysis

In order to observe the ability of clustering the arbitrary shape and different thresholds how to affect the clustering results, we test the DOCA-TWD algorithm again on the data set t10k, which has 10000 points from the Chameleon data sets [3]. The points in the same cluster marks in the same color and shape, and the points in different clusters marks different colors and shapes.

According to the rule (P) in Subsect. 3.1, if we keep α unchanged, the positive region of a cluster will not change. In other words, the main clustering structures don't change. If there is only the β changed, there is only the boundary regions changed; which will reflect the different clustering results under different

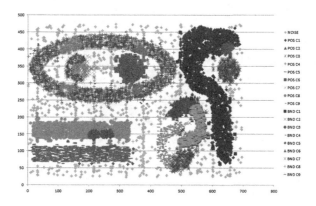

Fig. 3. $\alpha = 1.80$ $\beta = 0.02$

granularity. Thus, we run Algorithm 2 in the different granularity, namely, the threshold α always is 1.80 and β is set from 0.02 to 0.10. The results are recorded in Figs. 3, 4, 5 and 6, respectively. Because the result when $\alpha = 1.80$ $\beta = 0.06$ is very similar to the result when $\alpha = 1.80$ $\beta = 0.04$, we only show up one result here.

Observe Figs. 3, 4, 5 and 6, it's obvious that the positive regions don't change, and the boundary regions narrowed when β becomes bigger. In Fig. 3, C_4 and C_7, C_2 and C_9 have a large overlapping area in their boundary region, respectively. It's reasonable that those clusters should be merged into a more bigger granularity. That means the DOCA-TWD can discover the structure of the data set on a certain granularity. How to find the different clustering results autonomously under dynamic granularity is our further work.

After discussing the β how to effect clustering results, let's observe the α how to effect clustering results. According to the above discussions, clustering results will change when α changed. For example, when α is set to be an extreme small

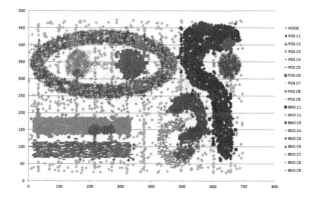

Fig. 4. $\alpha = 1.80$ $\beta = 0.04$

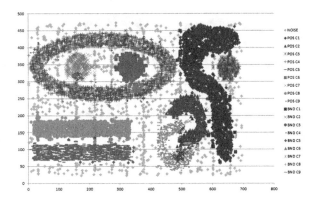

Fig. 5. $\alpha = 1.80\ \beta = 0.08$

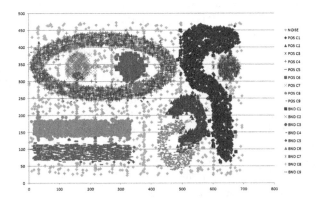

Fig. 6. $\alpha = 1.80\ \beta = 0.10$

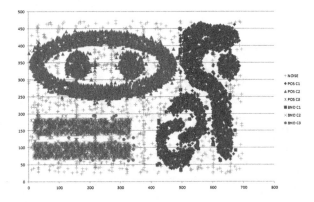

Fig. 7. $\alpha = 0.01\ \beta = 0.01$

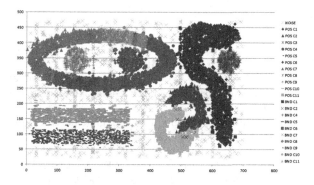

Fig. 8. $\alpha = 2.60$ $\beta = 2.00$

value, does there is only one cluster? Thus, we test on the t10k again under different α values. Figures 7 and 8 represent two results for saving space.

In Fig. 8, the number of clusters is eleven, while the number of clusters is 3 in Fig. 7. The conditions of merging for $\alpha = 2.60$ is much stricter than those for $\alpha = 0.01$. Anyway, Algorithm 2 still find out most of the clusters correct. This is an interesting result, even if we set α and β to be a very small value as 0.01, the Algorithm 2 does not put all the nodes into one cluster but still find three main clusters and recognizes the noise points correctly, which is shown in Fig. 7. The experimental results show that the proposed method doesn't depend too much on the thresholds.

6 Conclusion

In many applications such as network structure analysis, wireless sensor networks and biological information, an object should belong to more than one cluster, and as a result, cluster boundaries necessarily overlap. Three-way decisions rules constructed from the decision-theoretic rough set model are associated with different regions. This paper provides a three-way decision approach for density-based overlapping clustering. Here, each cluster is described by an interval set that is defined by a pair of sets called the lower and upper bounds. In addition, a two step density-based clustering algorithm is proposed and tested by using the new approach, where the first step is to get an initial clustering results based on the framework of density-based clustering, and the second step is based on the three-way decisions to obtain the final results, the positive region reflects the cluster structures and the boundary region describes the relationships between clusters. The experimental results indicate that the three-way decisions approach is effective to overlapping clustering, and doesn't depend too much on the thresholds.

Acknowledgments. This work was supported in part by the China NSFC grant (No.61379114 & No.61272060).

References

1. Asharaf, S., Murty, M.N.: An adaptive rough fuzzy single pass algorithm for clustering large data sets. Pattern Recogn. **36**(12), 3015–3018 (2003)
2. Aydin, N., Naït-Abdesselam, F., Pryyma, V., Turgut, D.: Overlapping clusters algorithm in ad hoc networks. In: 2010 IEEE Global Telecommunications Conference, pp. 1–5 (2010)
3. Chameleon data sets: http://glaros.dtc.umn.edu/gkhome/cluto/download
4. Chen, M., Miao, D.Q.: Interval set clustering. Expert Syst. Appl. **38**, 2923–2932 (2011)
5. Duan, L., Xu, L.D., Guo, F., Lee, J., Yan, B.P.: A local-density based spatial clustering algorithm with noise. Inf. Syst. **32**(7), 978–986 (2007)
6. Ester, M., Kriegel, H.P., Sander, J., Xu, X.W.: A density-based algorithm for discovering clusters in large spatial databases with noise. In: KDD (1996)
7. Fu, Q., Banerjee, A.: Multiplicative mixture models for overlapping clustering. In: IEEE International Conference on Data Mining, pp. 791–797 (2003)
8. Herbert, J.P., Yao, J.T.: Learning optimal parameters in decision-theoretic rough sets. In: Wen, P., Li, Y., Polkowski, L., Yao, Y., Tsumoto, S., Wang, G. (eds.) RSKT 2009. LNCS (LNAI), vol. 5589, pp. 610–617. Springer, Heidelberg (2009)
9. Lingras, P., Bhalchandra, P., Khamitkar, S., Mekewad, S., Rathod, R.: Crisp and soft clustering of mobile calls. In: Sombattheera, C., Agarwal, A., Udgata, S.K., Lavangnananda, K. (eds.) MIWAI 2011. LNCS (LNAI), vol. 7080, pp. 147–158. Springer, Heidelberg (2011)
10. Lingras, P., Yao, Y.Y.: Time complexity of rough clustering: GAs versus K-means. In: Alpigini, J.J., Peters, J.F., Skowron, A., Zhong, N. (eds.) RSCTC 2002. LNCS (LNAI), vol. 2475, pp. 263–270. Springer, Heidelberg (2002)
11. Liu, D., Yao, Y.Y., Li, T.R.: Three-way investment decisions with decision-theoretic rough sets. Int. J. Comput. Intell. Syst. **4**(1), 66–74 (2011)
12. Ma, S., Wang, T.J., Tang, S.W., Yang, D.Q., Gao, J.: A fast clustering algorithm based on reference and density. J. Softw. **14**(6), 1089–1095 (2003)
13. Obadi, G., Dráždilová, P., Hlaváček, L., Martinovič, J., Snášel, V.: A tolerance rough set based overlapping clustering for the DBLP data. In: IEEE/WIC/ACM International Conference on Web Intelligence and Intelligent Agent Technology, pp. 57–60 (2010)
14. Pawlak, Z.: Rough sets. Int. J. Comput. Inf. Sci. **11**(5), 341–356 (1982)
15. Pawlak, Z.: Rough classification. Int. J. Man-Mach. Stud. **20**(5), 469–483 (1984)
16. Ren, Y., Liu, X.D., Liu, W.Q.: DBCAMM: a novel density based clustering algorithm via using the Mahalanobis metric. Appl. Soft Comput. **12**(5), 1542–1554 (2010)
17. Sander, J., Ester, M., Kriegel, H.P., Xu, X.W.: Density-based clustering in spatial databases: the algorithm GDBSCAN and its applications. Data Min. Knowl. Disc. **2**(2), 169–194 (1998)
18. Subramaniam, S., Palpanas, T., Papadopoulos, D., Kalogeraki, V., Gunopulos, D.: Online outlier detection in sensor data using non-parametric models. In: Proceedings of the 32nd International Conference on Very Large Data Bases, pp. 187–198 (2006)
19. Takaki, M.: A extraction method of overlapping cluster based on network structure analysis. In: IEEE/WIC/ACM International Conferences on Web Intelligence and Intelligent Agent Technology, pp. 212–217 (2007)

20. Tsai, C.-F., Liu, C.-W.: KIDBSCAN: a new efficient data clustering algorithm. In: Rutkowski, L., Tadeusiewicz, R., Zadeh, L.A., Żurada, J.M. (eds.) ICAISC 2006. LNCS (LNAI), vol. 4029, pp. 702–711. Springer, Heidelberg (2006)
21. Weller, A.C.: Editorial Peer Review: Its Strengths & Weaknesses. Information Today Inc., Medford (2001)
22. Wu, X.H., Zhou, J.J.: Possibilistic fuzzy c-means clustering model using kernel methods. In: Proceeding of the 2005 International Conference on Computational Intelligence for Modelling, Control and Automation, and International Conference on Intelligent Agents, Web Technologies and Internet Commerce (CIMCA-IAWTIC'05), vol. 2, pp. 465–470 (2005)
23. Yao, Y.Y.: Three-way decisions with probabilistic rough sets. Inf. Sci. **180**(3), 341–353 (2010)
24. Yao, Y.Y.: The superiority of three-way decisions in probabilistic rough set models. Inf. Sci. **181**(6), 1080–1096 (2011)
25. Yao, Y.Y., Lingras, P., Wang, R.Z., Miao, D.Q.: Interval set cluster analysis: a reformulation. In: Sakai, H., Chakraborty, M.K., Hassanien, A.E., Ślęzak, D., Zhu, W. (eds.) RSFDGrC 2009. LNCS (LNAI), vol. 5908, pp. 398–405. Springer, Heidelberg (2009)
26. Yao, Y.Y., Wong, S.K.M.: A decision theoretic framework for approximating concepts. Int. J. Man-Mach. Stud. **37**(6), 793–809 (1992)
27. Yousri, N.A., Kamel, M.S., Ismail, M.A.: A possibilistic density based clustering for discovering clusters of arbitrary shapes and densities in high dimensional data. In: Huang, T., Zeng, Z., Li, C., Leung, C.S. (eds.) ICONIP 2012, Part III. LNCS, vol. 7665, pp. 577–584. Springer, Heidelberg (2012)
28. Yu, H., Luo, H.: A novel possibilistic fuzzy leader clustering algorithm. Int. J. Hybrid Intell. Syst. **8**(1), 31–40 (2011)
29. Zhou, B., Yao, Y.Y., Luo, J.G.: A three-way decision approach to email spam filtering. In: Farzindar, A., Keselj, V. (eds.) Canadian AI 2010. LNCS (LNAI), vol. 6085, pp. 28–39. Springer, Heidelberg (2010)
30. Lusseau, D., Newman, M.E.J.: Identifying the role that animals play in their social networks. Proc. R. Soc. Lond. Ser. B Biol. Sci. **271**(Suppl 6), S477–S481 (2004)

Three-Way Decisions in Stochastic Decision-Theoretic Rough Sets

Dun Liu[1]([⊠]), Tianrui Li[2], and Decui Liang[3]

[1] School of Economics and Management, Southwest Jiaotong University,
Chengdu 610031, People's Republic of China
newton83@163.com
[2] School of Information Science and Technology, Southwest Jiaotong University,
Chengdu 610031, People's Republic of China
trli@swjtu.edu.cn
[3] School of Management and Economics, University of Electronic Science
and Technology of China, Chengdu 610054, People's Republic of China
decuiliang@126.com

Abstract. In the previous decision-theoretic rough sets (DTRS), its loss function values are precise. This paper extends the precise values of loss functions to a more realistic stochastic environment. The stochastic loss functions are induced to decision-theoretic rough set theory based on the bayesian decision theory. A model of stochastic decision-theoretic rough set theory (SDTRS) is built with respect to the minimum bayesian expected risk. The corresponding propositions and criteria of SDTRS are also analyzed. Furthermore, we investigate two special SDTRS models under the uniform distribution and the normal distribution, respectively. Finally, an empirical study of Public-Private Partnerships (PPP) project investment validates the reasonability and effectiveness of the proposed models.

Keywords: Decision-theoretic rough sets · Confidence interval · Statistic distribution · Stochastic

1 Introduction

Three-way decisions, a new viewpoint to investigate probabilistic rough sets, has become more and more important on applications of rough sets. Different from two-way decisions (acceptance and rejection), three-way decisions utilize the third way, *i.e.*, deferment, to deal with the things which cannot decide immediately at the moment. With respect to the three regions (positive region, boundary region and negative region), Yao introduced the notion of three-way decisions and provided an interpretation of rules in rough sets, consisting of the positive, boundary and negative rules [49,50,52,54,55]. Apparently, positive rules, negative rules and boundary rules correspond to making decisions of acceptance, rejection and deferring a definite decision, respectively. These ideas of three-way decisions have in fact been applied to many fields, both in theoretical analysis

© Springer-Verlag Berlin Heidelberg 2014
J.F. Peters et al. (Eds.): Transactions on Rough Sets XVIII, LNCS 8449, pp. 110–130, 2014.
DOI: 10.1007/978-3-662-44680-5_7

[8,10,15–19,41] and applications, *e.g.*, medical clinic [31,42], products inspecting process [39], environmental management [7], data packs selection [38], E-learning [2], email spam filtering [57,59], oil investment [16,40] and government decisions [21].

Intuitively, probabilistic rough sets can handle three-way decisions by using two thresholds. Compared with Pawlak rough sets, probabilistic rough sets consider some certain levels of tolerance for errors and utilize two parameters α and β to divide the universe into three parts [33,34,45,47,61,62]. The rules generated from these three parts lead to three-way decisions. However, as pointed out by Liu et al. in [24], one big challenge of probabilistic rough sets is to estimate the values of α and β. In most probabilistic rough set models, the two parameters are given by experts with intuitive experience estimating [15]. Observed by this issue, Yao et al. introduced Bayesian decision procedure to rough sets, and proposed Decision-theoretic rough set model (DTRSM) [43,44,46]. In DTRS, the two parameters α and β can be directly and systematically calculated by minimizing the decision costs, which gives a brief semantics explanation in practical applications with minimum decision risks. However, the loss functions in DTRS are assumed as precise numerical values, the decision makers may hardly estimate the precise loss function values, especially when the decision problem is complex [20]. By considering the domain knowledge of decision maker is limited and vague, Liu et al. suggested to use some uncertain information (*i.e.* stochastic, vague or rough information) instead of precise one in real decision procedure [20]. For example, under the vague and rough environments, Liu et al. introduced interval-valued number to DTRS, and proposed interval-valued decision-theoretic rough sets [22]. Liu et al. further introduced fuzzy loss functions and fuzzy interval loss functions to DTRS, then proposed fuzzy decision-theoretic rough sets [23] and fuzzy interval decision-theoretic rough sets [25], respectively. Liang et al. [12] generalized a concept of the precise value of loss function to triangular fuzzy number, and proposed triangular fuzzy decision-theoretic rough sets (TFDTRS). Obviously, the extensions of loss functions in [22,23,25] can handle one kind of uncertainty during the real decision process. Furthermore, as one important description of uncertainty, stochastic number is used to indicate that a particular subject is seen from point of view of randomness, which refers to systems whose behavior is intrinsically non-deterministic, sporadic and categorically not intermittent. In this paper, we use stochastic number to replace precise number, and discuss the stochastic extension of decision-theoretic rough sets model.

The remainder of this paper is organized as follows: Sect. 2 provides the basic concepts of stochastic model and DTRS. In Sect. 3, we introduce the stochastic loss functions to DTRS. A new model of stochastic decision-theoretic rough sets (SDTRS) is proposed. Two special SDTRS models under uniform distribution and normal distribution are further investigated. Then, a case study of PPP project investment problem is given to illustrate our approach in Sect. 4. Section 5 concludes the paper and outlines the future work.

2   Preliminaries

Basic concepts, notations and results of stochastic model and DTRS are briefly reviewed in this section [8,10,15–18,44,50,52,53].

2.1   Stochastic Model and Confidence Interval

Stochastic is synonymous with "random". It is used to indicate that a particular subject is seen from point of view of randomness. Stochastic is often used as counterpart of the word "deterministic", which means that random phenomena are not involved. Therefore, stochastic models are based on random trials. In probability theory, the subsequent state of a stochastic system is determined by the system's predictable actions and a random element.

Definition 1. *In the basic statistical model, an observable random variable X taking values in a set S. In general, X can have quite a complicated structure. For example, if the experiment is to sample n objects from a population and record various measurements of interest, then*

$$X = (X_1, X_2, \ldots, X_n). \tag{1}$$

where X_i is the vector of measurements for the ith object. Suppose $(X_1, X_2, , X_n)$ are independent and identically distributed, we have a random sample of size n from the common distribution. Suppose that the distribution of X depends on a parameter θ taking values in a parameter space Θ. The parameter may also be vector-valued, in which case $\Theta \subseteq R^k$ for some $k \in N_+$ and $\theta = (\theta_1, \theta_2, \ldots, \theta_k)$.

A confidence interval (CI) is a type of interval estimate of a population parameter and is used to indicate the reliability of an estimate. It is an observed interval, which contains the parameter determined by the confidence level or confidence coefficient. In statistics, CI is a type of interval estimate of a population parameter and is used to indicate the reliability of an estimate.

Definition 2. *Let X be a random sample from a probability distribution with statistical parameters θ, which is a quantity to be estimated. A confidence interval for the parameter θ, with confidence level or confidence coefficient φ, is an interval with random endpoints $(u(X), v(X))$, determined by the pair of random variables $u(X)$ and $v(X)$, with the property:*

$$Pr_\theta(u(X) < \theta < v(X)) = \varphi \ for \ all \ \theta. \tag{2}$$

The number l, with typical values close to but not greater than 1, is sometimes given in the form $1 - l$ (l denotes significance level with $0 < l < 1$). Pr_θ indicates the probability distribution of X characterized by θ. An important part of this specification is that the random interval $(u(X), v(X))$ covers the unknown value θ with a high probability no matter what the true value of θ actually is. Hence, a confidence interval is a statement like "θ is between 5 and 10 with probability 95 %."

2.2   Decision-Theoretic Rough Sets

For the Bayesian decision procedure, the DTRS model is composed of 2 states and 3 actions. The set of states is given by $\Omega = \{C, \neg C\}$ indicating that an object is in C and not in C, respectively. The set of actions is given by $\mathcal{A} = \{a_P, a_B, a_N\}$, where a_P, a_B, and a_N represent the three actions in classifying an object x, namely, deciding $x \in \text{POS}(C)$, deciding $x \in \text{BND}(C)$, and deciding $x \in \text{NEG}(C)$, respectively. The loss function λ regarding the risk or cost of actions in different states is given by the 3×2 matrix:

	C (P)	$\neg C$ (N)
a_P	λ_{PP}	λ_{PN}
a_B	λ_{BP}	λ_{BN}
a_N	λ_{NP}	λ_{NN}

In the matrix, λ_{PP}, λ_{BP} and λ_{NP} denote the losses incurred for taking actions of a_P, a_B and a_N, respectively, when an object belongs to C. Similarly, λ_{PN}, λ_{BN} and λ_{NN} denote the losses incurred for taking the same actions when the object belongs to $\neg C$. $Pr(C|[x])$ is the conditional probability of an object x belonging to C given that the object is described by its equivalence class $[x]$. For an object x, the expected loss $R(a_i|[x])$ associated with taking the individual actions can be expressed as:

$$R(a_P|[x]) = \lambda_{PP}Pr(C|[x]) + \lambda_{PN}Pr(\neg C|[x]),$$
$$R(a_B|[x]) = \lambda_{BP}Pr(C|[x]) + \lambda_{BN}Pr(\neg C|[x]),$$
$$R(a_N|[x]) = \lambda_{NP}Pr(C|[x]) + \lambda_{NN}Pr(\neg C|[x]). \tag{3}$$

The Bayesian decision procedure suggests the following minimum-cost decision rules:

(P) If $R(a_P|[x]) \leq R(a_B|[x])$ and $R(a_P|[x]) \leq R(a_N|[x])$, decide $x \in \text{POS}(C)$;

(B) If $R(a_B|[x]) \leq R(a_P|[x])$ and $R(a_B|[x]) \leq R(a_N|[x])$, decide $x \in \text{BND}(C)$;

(N) If $R(a_N|[x]) \leq R(a_P|[x])$ and $R(a_N|[x]) \leq R(a_B|[x])$, decide $x \in \text{NEG}(C)$.

Since $Pr(C|[x]) + Pr(\neg C|[x]) = 1$, we simplify the rules based only on the probability $Pr(C|[x])$ and the loss function.

Furthermore, by considering the fact the loss of classifying an object x belonging to C into the positive region $POS(C)$ is less than or equal to the loss of classifying x into the boundary region $BND(C)$, and both of these losses are strictly less than the loss of classifying x into the negative region $NEG(C)$. The reverse order of losses is used for classifying an object not in C. The reverse order of loss is hold for classifying an object x is not in C. Hence, the loss functions may satisfy:

$$\lambda_{PP} \leq \lambda_{BP} < \lambda_{NP};$$
$$\lambda_{NN} \leq \lambda_{BN} < \lambda_{PN}. \tag{4}$$

The decision rules (P)-(N) can be expressed concisely as:

(P) If $Pr(C|[x]) \geq \alpha$ and $Pr(C|[x]) \geq \gamma$, decide $x \in \text{POS}(C)$;

(B) If $Pr(C|[x]) \leq \alpha$ and $Pr(C|[x]) \geq \beta$, decide $x \in \text{BND}(C)$;

(N) If $Pr(C|[x]) \leq \beta$ and $Pr(C|[x]) \leq \gamma$, decide $x \in \text{NEG}(C)$.

The threshold values α, β, γ are given by:

$$\alpha = \frac{(\lambda_{PN} - \lambda_{BN})}{(\lambda_{PN} - \lambda_{BN}) + (\lambda_{BP} - \lambda_{PP})},$$

$$\beta = \frac{(\lambda_{BN} - \lambda_{NN})}{(\lambda_{BN} - \lambda_{NN}) + (\lambda_{NP} - \lambda_{BP})},$$

$$\gamma = \frac{(\lambda_{PN} - \lambda_{NN})}{(\lambda_{PN} - \lambda_{NN}) + (\lambda_{NP} - \lambda_{PP})}. \tag{5}$$

In addition, as a well-defined boundary region, the conditions of rule (B) suggest that $\alpha > \beta$, that is, $\frac{(\lambda_{PN}-\lambda_{BN})}{(\lambda_{PN}-\lambda_{BN})+(\lambda_{BP}-\lambda_{PP})} > \frac{(\lambda_{BN}-\lambda_{NN})}{(\lambda_{BN}-\lambda_{NN})+(\lambda_{NP}-\lambda_{BP})}$. Hence, we obtain: $\frac{\lambda_{NP}-\lambda_{BP}}{\lambda_{BN}-\lambda_{NN}} > \frac{\lambda_{BP}-\lambda_{PP}}{\lambda_{PN}-\lambda_{BN}}$. Because of the inequality $\frac{b}{a} > \frac{d}{c} \implies \frac{b}{a} > \frac{b+d}{a+c} > \frac{d}{c}, (a, b, c, d > 0)$, we have: $0 \leq \beta < \gamma < \alpha \leq 1$. In this case, after tie-breaking, the following simplified rules are obtained:

(P1)   If $Pr(C|[x]) \geq \alpha$, decide $x \in \text{POS}(C)$;

(B1)   If $\beta < Pr(C|[x]) < \alpha$, decide $x \in \text{BND}(C)$;

(N1)   If $Pr(C|[x]) \leq \beta$, decide $x \in \text{NEG}(C)$.

3   Extension of Decision-Theoretic Rough Set Models

In this section, we discuss the extension of decision-theoretic rough set models when the loss function is stochastic number. In Sect. 3.1, we introduce CI to estimate the stochastic loss functions and propose stochastic decision-theoretic rough sets (SDTRS). In Sect. 3.2, we investigate two special cases, the SDTRS models with the uniform distribution and the normal distribution, respectively.

3.1   Stochastic Decision-Theoretic Rough Sets

As an extension of DTRS, SDTRS utilizes six stochastic loss functions $\lambda_{\bullet\bullet}^{\varepsilon}$ ($\bullet = P, B, N$, ε denotes the loss function $\lambda_{\bullet\bullet}^{\varepsilon}$ is a stochastic number) to describe the cost of actions in different states, which is given by the 3×2 matrix:

	C (P)	$\neg C$ (N)
a_P	$\lambda_{PP}^{\varepsilon}$	$\lambda_{PN}^{\varepsilon}$
a_B	$\lambda_{BP}^{\varepsilon}$	$\lambda_{BN}^{\varepsilon}$
a_N	$\lambda_{NP}^{\varepsilon}$	$\lambda_{NN}^{\varepsilon}$

In the matrix, the estimation of $\lambda^\varepsilon_{\bullet\bullet}$ becomes very important. Because of CI can give a range of plausible values for the estimate of the unknown population parameter, we introduce CI to SDTRS to estimate the loss function $\lambda^\varepsilon_{\bullet\bullet}$. As we know, the identification of the population parameter with CI depends on four factors: the sample mean, the standard deviation, the sample size and the population distributions. In the following, we investigate two scenarios for estimating $\lambda^\varepsilon_{\bullet\bullet}$ with CI.

3.1.1 Confidence Intervals of $\lambda^\varepsilon_{\bullet\bullet}$ (σ Known)

In this scenario, suppose n denotes the size of random sample, $\mu^\varepsilon_{\bullet\bullet}$ denotes the sample mean, and $\sigma^\varepsilon_{\bullet\bullet}$ denotes the population standard deviation. If the population of $\lambda^\varepsilon_{\bullet\bullet}$ is normally distributed or $n \geq 30$. Given a significance level l, the estimation of $\lambda^\varepsilon_{\bullet\bullet}$ can be calculated as:

$$\mu^\varepsilon_{\bullet\bullet} - Z_{\frac{l}{2}} \cdot \frac{\sigma^\varepsilon_{\bullet\bullet}}{\sqrt{n}} \leq \lambda^\varepsilon_{\bullet\bullet} \leq \mu^\varepsilon_{\bullet\bullet} + Z_{\frac{l}{2}} \cdot \frac{\sigma^\varepsilon_{\bullet\bullet}}{\sqrt{n}}. \tag{6}$$

In (6), there are two steps to find the required Z critical value: (i) Find $\frac{l}{2}$; (ii) Take this value and locate it in the standard normal probability table and identify the Z critical value. Therefore, one can use an interval $[\mu^\varepsilon_{\bullet\bullet} - Z_{\frac{l}{2}} \cdot \frac{\sigma^\varepsilon_{\bullet\bullet}}{\sqrt{n}}, \mu^\varepsilon_{\bullet\bullet} + Z_{\frac{l}{2}} \cdot \frac{\sigma^\varepsilon_{\bullet\bullet}}{\sqrt{n}}]$ to estimate $\lambda^\varepsilon_{\bullet\bullet}$ by choosing a proper l, and the 3×2 matrix can be expressed as:

	C (P)	$\neg C$ (N)
a_P	$[\mu^\varepsilon_{PP} - Z_{\frac{l}{2}} \cdot \frac{\sigma^\varepsilon_{PP}}{\sqrt{n}}, \mu^\varepsilon_{PP} + Z_{\frac{l}{2}} \cdot \frac{\sigma^\varepsilon_{PP}}{\sqrt{n}}]$	$[\mu^\varepsilon_{PN} - Z_{\frac{l}{2}} \cdot \frac{\sigma^\varepsilon_{PN}}{\sqrt{n}}, \mu^\varepsilon_{PN} + Z_{\frac{l}{2}} \cdot \frac{\sigma^\varepsilon_{PN}}{\sqrt{n}}]$
a_B	$[\mu^\varepsilon_{BP} - Z_{\frac{l}{2}} \cdot \frac{\sigma^\varepsilon_{BP}}{\sqrt{n}}, \mu^\varepsilon_{BP} + Z_{\frac{l}{2}} \cdot \frac{\sigma^\varepsilon_{BP}}{\sqrt{n}}]$	$[\mu^\varepsilon_{BN} - Z_{\frac{l}{2}} \cdot \frac{\sigma^\varepsilon_{BN}}{\sqrt{n}}, \mu^\varepsilon_{BN} + Z_{\frac{l}{2}} \cdot \frac{\sigma^\varepsilon_{BN}}{\sqrt{n}}]$
a_N	$[\mu^\varepsilon_{NP} - Z_{\frac{l}{2}} \cdot \frac{\sigma^\varepsilon_{NP}}{\sqrt{n}}, \mu^\varepsilon_{NP} + Z_{\frac{l}{2}} \cdot \frac{\sigma^\varepsilon_{NP}}{\sqrt{n}}]$	$[\mu^\varepsilon_{NN} - Z_{\frac{l}{2}} \cdot \frac{\sigma^\varepsilon_{NN}}{\sqrt{n}}, \mu^\varepsilon_{NN} + Z_{\frac{l}{2}} \cdot \frac{\sigma^\varepsilon_{NN}}{\sqrt{n}}]$

By considering the basic assumption of DTRS, it requires all the loss functions $\lambda^\varepsilon_{\bullet\bullet} \geq 0$, that is, $\mu^\varepsilon_{\bullet\bullet} - Z_{\frac{l}{2}} \cdot \frac{\sigma^\varepsilon_{\bullet\bullet}}{\sqrt{n}} \geq 0$.

For simplicity, we mainly consider two cases: the lower bound and the upper bound of $\lambda^\varepsilon_{\bullet\bullet}$. Following the discussions in (3), the expected losses with the lower bound can be expressed as:

$$R_L(a_P|[x]) = (\mu^\varepsilon_{PP} - Z_{\frac{l}{2}} \cdot \frac{\sigma^\varepsilon_{PP}}{\sqrt{n}}) \cdot Pr(C|[x]) + (\mu^\varepsilon_{PP} - Z_{\frac{l}{2}} \cdot \frac{\sigma^\varepsilon_{PN}}{\sqrt{n}}) \cdot Pr(\neg C|[x]),$$

$$R_L(a_B|[x]) = (\mu^\varepsilon_{BP} - Z_{\frac{l}{2}} \cdot \frac{\sigma^\varepsilon_{BP}}{\sqrt{n}}) \cdot Pr(C|[x]) + (\mu^\varepsilon_{BN} - Z_{\frac{l}{2}} \cdot \frac{\sigma^\varepsilon_{BN}}{\sqrt{n}}) \cdot Pr(\neg C|[x]),$$

$$R_L(a_N|[x]) = (\mu^\varepsilon_{NP} - Z_{\frac{l}{2}} \cdot \frac{\sigma^\varepsilon_{NP}}{\sqrt{n}}) \cdot Pr(C|[x]) + (\mu^\varepsilon_{NN} - Z_{\frac{l}{2}} \cdot \frac{\sigma^\varepsilon_{NN}}{\sqrt{n}}) \cdot Pr(\neg C|[x]).$$

Similarly, the expected losses with the upper bound can be expressed as:

$$R_U(a_P|[x]) = (\mu_{PP}^\varepsilon + Z_{\frac{L}{2}} \cdot \frac{\sigma_{PP}^\varepsilon}{\sqrt{n}}) \cdot Pr(C|[x]) + (\mu_{PP}^\varepsilon + Z_{\frac{L}{2}} \cdot \frac{\sigma_{PN}^\varepsilon}{\sqrt{n}}) \cdot Pr(\neg C|[x]),$$

$$R_U(a_B|[x]) = (\mu_{BP}^\varepsilon + Z_{\frac{L}{2}} \cdot \frac{\sigma_{BP}^\varepsilon}{\sqrt{n}}) \cdot Pr(C|[x]) + (\mu_{BN}^\varepsilon + Z_{\frac{L}{2}} \cdot \frac{\sigma_{BN}^\varepsilon}{\sqrt{n}}) \cdot Pr(\neg C|[x]),$$

$$R_U(a_N|[x]) = (\mu_{NP}^\varepsilon + Z_{\frac{L}{2}} \cdot \frac{\sigma_{NP}^\varepsilon}{\sqrt{n}}) \cdot Pr(C|[x]) + (\mu_{NN}^\varepsilon + Z_{\frac{L}{2}} \cdot \frac{\sigma_{NN}^\varepsilon}{\sqrt{n}}) \cdot Pr(\neg C|[x]).$$

The Bayesian decision procedure suggests the following minimum-cost decision rules:

(P$_*$) If $R_*(a_P|[x]) \leq R_*(a_B|[x])$ and $R_*(a_P|[x]) \leq R_*(a_N|[x])$, decide $x \in POS(C)$;

(B$_*$) If $R_*(a_B|[x]) \leq R_*(a_P|[x])$ and $R_*(a_B|[x]) \leq R_*(a_N|[x])$, decide $x \in BND(C)$;

(N$_*$) If $R_*(a_N|[x]) \leq R_*(a_P|[x])$ and $R_*(a_N|[x]) \leq R_*(a_B|[x])$, decide $x \in NEG(C)$.

where $* = L, U$.

On the basis of conditions reported in (4), the losses in this model satisfy the following constraints:

$$(\mu_{PP}^\varepsilon - Z_{\frac{L}{2}} \cdot \frac{\sigma_{PP}^\varepsilon}{\sqrt{n}}) \leq (\mu_{BP}^\varepsilon - Z_{\frac{L}{2}} \cdot \frac{\sigma_{BP}^\varepsilon}{\sqrt{n}}) < (\mu_{NP}^\varepsilon - Z_{\frac{L}{2}} \cdot \frac{\sigma_{NP}^\varepsilon}{\sqrt{n}});$$

$$(\mu_{NN}^\varepsilon - Z_{\frac{L}{2}} \cdot \frac{\sigma_{NN}^\varepsilon}{\sqrt{n}}) \leq (\mu_{BN}^\varepsilon - Z_{\frac{L}{2}} \cdot \frac{\sigma_{BN}^\varepsilon}{\sqrt{n}}) < (\mu_{PN}^\varepsilon - Z_{\frac{L}{2}} \cdot \frac{\sigma_{PN}^\varepsilon}{\sqrt{n}});$$

$$(\mu_{PP}^\varepsilon + Z_{\frac{L}{2}} \cdot \frac{\sigma_{PP}^\varepsilon}{\sqrt{n}}) \leq (\mu_{BP}^\varepsilon + Z_{\frac{L}{2}} \cdot \frac{\sigma_{BP}^\varepsilon}{\sqrt{n}}) < (\mu_{NP}^\varepsilon + Z_{\frac{L}{2}} \cdot \frac{\sigma_{NP}^\varepsilon}{\sqrt{n}});$$

$$(\mu_{NN}^\varepsilon + Z_{\frac{L}{2}} \cdot \frac{\sigma_{NN}^\varepsilon}{\sqrt{n}}) \leq (\mu_{BN}^\varepsilon + Z_{\frac{L}{2}} \cdot \frac{\sigma_{BN}^\varepsilon}{\sqrt{n}}) < (\mu_{PN}^\varepsilon + Z_{\frac{L}{2}} \cdot \frac{\sigma_{PN}^\varepsilon}{\sqrt{n}}). \tag{7}$$

The decision rules (P$_*$)-(N$_*$) can be expressed concisely as:

(P$_*$) If $Pr(C|[x]) \geq \alpha_*$ and $Pr(C|[x]) \geq \gamma_*$, decide $x \in POS(C)$;

(B$_*$) If $Pr(C|[x]) \leq \alpha_*$ and $Pr(C|[x]) \geq \beta_*$, decide $x \in BND(C)$;

(N$_*$) If $Pr(C|[x]) \leq \beta_*$ and $Pr(C|[x]) \leq \gamma_*$, decide $x \in NEG(C)$.

Since $Pr(C|[x]) + Pr(\neg C|[x]) = 1$, the threshold values under above two cases are calculated.

$$\alpha_L = \frac{[(\mu_{PN}^\varepsilon - Z_{\frac{L}{2}} \cdot \frac{\sigma_{PN}^\varepsilon}{\sqrt{n}}) - (\mu_{BN}^\varepsilon - Z_{\frac{L}{2}} \cdot \frac{\sigma_{BN}^\varepsilon}{\sqrt{n}})]}{[(\mu_{PN}^\varepsilon - Z_{\frac{L}{2}} \cdot \frac{\sigma_{PN}^\varepsilon}{\sqrt{n}}) - (\mu_{BN}^\varepsilon - Z_{\frac{L}{2}} \cdot \frac{\sigma_{BN}^\varepsilon}{\sqrt{n}})] + [(\mu_{BP}^\varepsilon - Z_{\frac{L}{2}} \cdot \frac{\sigma_{BP}^\varepsilon}{\sqrt{n}}) - (\mu_{PP}^\varepsilon - Z_{\frac{L}{2}} \cdot \frac{\sigma_{PP}^\varepsilon}{\sqrt{n}})]},$$

$$\beta_L = \frac{[(\mu_{BN}^\varepsilon - Z_{\frac{L}{2}} \cdot \frac{\sigma_{BN}^\varepsilon}{\sqrt{n}}) - (\mu_{NN}^\varepsilon - Z_{\frac{L}{2}} \cdot \frac{\sigma_{NN}^\varepsilon}{\sqrt{n}})]}{[(\mu_{BN}^\varepsilon - Z_{\frac{L}{2}} \cdot \frac{\sigma_{BN}^\varepsilon}{\sqrt{n}}) - (\mu_{NN}^\varepsilon - Z_{\frac{L}{2}} \cdot \frac{\sigma_{NN}^\varepsilon}{\sqrt{n}})] + [(\mu_{NP}^\varepsilon - Z_{\frac{L}{2}} \cdot \frac{\sigma_{NP}^\varepsilon}{\sqrt{n}}) - (\mu_{BP}^\varepsilon - Z_{\frac{L}{2}} \cdot \frac{\sigma_{BP}^\varepsilon}{\sqrt{n}})]},$$

$$\gamma_L = \frac{[(\mu_{PN}^\varepsilon - Z_{\frac{L}{2}} \cdot \frac{\sigma_{PN}^\varepsilon}{\sqrt{n}}) - (\mu_{NN}^\varepsilon - Z_{\frac{L}{2}} \cdot \frac{\sigma_{NN}^\varepsilon}{\sqrt{n}})]}{[(\mu_{PN}^\varepsilon - Z_{\frac{L}{2}} \cdot \frac{\sigma_{PN}^\varepsilon}{\sqrt{n}}) - (\mu_{NN}^\varepsilon - Z_{\frac{L}{2}} \cdot \frac{\sigma_{NN}^\varepsilon}{\sqrt{n}})] + [(\mu_{NP}^\varepsilon - Z_{\frac{L}{2}} \cdot \frac{\sigma_{NP}^\varepsilon}{\sqrt{n}}) - (\mu_{PP}^\varepsilon - Z_{\frac{L}{2}} \cdot \frac{\sigma_{PP}^\varepsilon}{\sqrt{n}})]}.$$

$$\alpha_U = \frac{[(\mu_{PN}^{\varepsilon} + Z_{\frac{l}{2}} \cdot \frac{\sigma_{PN}^{\varepsilon}}{\sqrt{n}}) - (\mu_{BN}^{\varepsilon} + Z_{\frac{l}{2}} \cdot \frac{\sigma_{BN}^{\varepsilon}}{\sqrt{n}})]}{[(\mu_{PN}^{\varepsilon} + Z_{\frac{l}{2}} \cdot \frac{\sigma_{PN}^{\varepsilon}}{\sqrt{n}}) - (\mu_{BN}^{\varepsilon} + Z_{\frac{l}{2}} \cdot \frac{\sigma_{BN}^{\varepsilon}}{\sqrt{n}})] + [(\mu_{BP}^{\varepsilon} + Z_{\frac{l}{2}} \cdot \frac{\sigma_{BP}^{\varepsilon}}{\sqrt{n}}) - (\mu_{PP}^{\varepsilon} + Z_{\frac{l}{2}} \cdot \frac{\sigma_{PP}^{\varepsilon}}{\sqrt{n}})]},$$

$$\beta_U = \frac{[(\mu_{BN}^{\varepsilon} + Z_{\frac{l}{2}} \cdot \frac{\sigma_{BN}^{\varepsilon}}{\sqrt{n}}) - (\mu_{NN}^{\varepsilon} + Z_{\frac{l}{2}} \cdot \frac{\sigma_{NN}^{\varepsilon}}{\sqrt{n}})]}{[(\mu_{BN}^{\varepsilon} + Z_{\frac{l}{2}} \cdot \frac{\sigma_{BN}^{\varepsilon}}{\sqrt{n}}) - (\mu_{NN}^{\varepsilon} + Z_{\frac{l}{2}} \cdot \frac{\sigma_{NN}^{\varepsilon}}{\sqrt{n}})] + [(\mu_{NP}^{\varepsilon} + Z_{\frac{l}{2}} \cdot \frac{\sigma_{NP}^{\varepsilon}}{\sqrt{n}}) - (\mu_{BP}^{\varepsilon} + Z_{\frac{l}{2}} \cdot \frac{\sigma_{BP}^{\varepsilon}}{\sqrt{n}})]},$$

$$\gamma_U = \frac{[(\mu_{PN}^{\varepsilon} + Z_{\frac{l}{2}} \cdot \frac{\sigma_{PN}^{\varepsilon}}{\sqrt{n}}) - (\mu_{NN}^{\varepsilon} + Z_{\frac{l}{2}} \cdot \frac{\sigma_{NN}^{\varepsilon}}{\sqrt{n}})]}{[(\mu_{PN}^{\varepsilon} + Z_{\frac{l}{2}} \cdot \frac{\sigma_{PN}^{\varepsilon}}{\sqrt{n}}) - (\mu_{NN}^{\varepsilon} + Z_{\frac{l}{2}} \cdot \frac{\sigma_{NN}^{\varepsilon}}{\sqrt{n}})] + [(\mu_{NP}^{\varepsilon} + Z_{\frac{l}{2}} \cdot \frac{\sigma_{NP}^{\varepsilon}}{\sqrt{n}}) - (\mu_{PP}^{\varepsilon} + Z_{\frac{l}{2}} \cdot \frac{\sigma_{PP}^{\varepsilon}}{\sqrt{n}})]}.$$

The conditions of rule (B_*) suggest that $\alpha_* > \beta_*$, we have:

$$\frac{[(\mu_{NP}^{\varepsilon} - Z_{\frac{l}{2}} \cdot \frac{\sigma_{NP}^{\varepsilon}}{\sqrt{n}}) - (\mu_{BP}^{\varepsilon} - Z_{\frac{l}{2}} \cdot \frac{\sigma_{BP}^{\varepsilon}}{\sqrt{n}})]}{[(\mu_{BN}^{\varepsilon} - Z_{\frac{l}{2}} \cdot \frac{\sigma_{BN}^{\varepsilon}}{\sqrt{n}}) - (\mu_{NN}^{\varepsilon} - Z_{\frac{l}{2}} \cdot \frac{\sigma_{NN}^{\varepsilon}}{\sqrt{n}})]} > \frac{[(\mu_{BP}^{\varepsilon} - Z_{\frac{l}{2}} \cdot \frac{\sigma_{BP}^{\varepsilon}}{\sqrt{n}}) - (\mu_{PP}^{\varepsilon} - Z_{\frac{l}{2}} \cdot \frac{\sigma_{PP}^{\varepsilon}}{\sqrt{n}})]}{[(\mu_{PN}^{\varepsilon} - Z_{\frac{l}{2}} \cdot \frac{\sigma_{PN}^{\varepsilon}}{\sqrt{n}}) - (\mu_{BN}^{\varepsilon} - Z_{\frac{l}{2}} \cdot \frac{\sigma_{BN}^{\varepsilon}}{\sqrt{n}})]},$$

$$\frac{[(\mu_{NP}^{\varepsilon} + Z_{\frac{l}{2}} \cdot \frac{\sigma_{NP}^{\varepsilon}}{\sqrt{n}}) - (\mu_{BP}^{\varepsilon} + Z_{\frac{l}{2}} \cdot \frac{\sigma_{BP}^{\varepsilon}}{\sqrt{n}})]}{[(\mu_{BN}^{\varepsilon} + Z_{\frac{l}{2}} \cdot \frac{\sigma_{BN}^{\varepsilon}}{\sqrt{n}}) - (\mu_{NN}^{\varepsilon} + Z_{\frac{l}{2}} \cdot \frac{\sigma_{NN}^{\varepsilon}}{\sqrt{n}})]} > \frac{[(\mu_{BP}^{\varepsilon} + Z_{\frac{l}{2}} \cdot \frac{\sigma_{BP}^{\varepsilon}}{\sqrt{n}}) - (\mu_{PP}^{\varepsilon} + Z_{\frac{l}{2}} \cdot \frac{\sigma_{PP}^{\varepsilon}}{\sqrt{n}})]}{[(\mu_{PN}^{\varepsilon} + Z_{\frac{l}{2}} \cdot \frac{\sigma_{PN}^{\varepsilon}}{\sqrt{n}}) - (\mu_{BN}^{\varepsilon} + Z_{\frac{l}{2}} \cdot \frac{\sigma_{BN}^{\varepsilon}}{\sqrt{n}})]}.$$

Following our discussions in Sect. 2.3, we can easily get: $0 \le \beta_* < \gamma_* < \alpha_* \le 1$, $* = L, U$. The decision rules (P_*)-(N_*) can be simplified as follows:

(P1$_*$)   If $Pr(C|[x]) \ge \alpha_*$, decide $x \in \text{POS}(C)$;
(B1$_*$)   If $\beta_* < Pr(C|[x]) < \alpha_*$, decide $x \in \text{BND}(C)$;
(N1$_*$)   If $Pr(C|[x]) \le \beta_*$, decide $x \in \text{NEG}(C)$.

where $* = L, U$. Hence, we can determine the results by comparing the three threshold values.

Furthermore, we consider one type of stricter condition based on (7).

$$(\mu_{PP}^{\varepsilon} - Z_{\frac{l}{2}} \cdot \frac{\sigma_{PP}^{\varepsilon}}{\sqrt{n}}) < (\mu_{PP}^{\varepsilon} + Z_{\frac{l}{2}} \cdot \frac{\sigma_{PP}^{\varepsilon}}{\sqrt{n}}) \le (\mu_{BP}^{\varepsilon} - Z_{\frac{l}{2}} \cdot \frac{\sigma_{BP}^{\varepsilon}}{\sqrt{n}})$$

$$< (\mu_{BP}^{\varepsilon} + Z_{\frac{l}{2}} \cdot \frac{\sigma_{BP}^{\varepsilon}}{\sqrt{n}}) < (\mu_{NP}^{\varepsilon} - Z_{\frac{l}{2}} \cdot \frac{\sigma_{NP}^{\varepsilon}}{\sqrt{n}}) < (\mu_{NP}^{\varepsilon} + Z_{\frac{l}{2}} \cdot \frac{\sigma_{NP}^{\varepsilon}}{\sqrt{n}});$$

$$(\mu_{NN}^{\varepsilon} - Z_{\frac{l}{2}} \cdot \frac{\sigma_{NN}^{\varepsilon}}{\sqrt{n}}) < (\mu_{NN}^{\varepsilon} + Z_{\frac{l}{2}} \cdot \frac{\sigma_{NN}^{\varepsilon}}{\sqrt{n}}) \le (\mu_{BN}^{\varepsilon} - Z_{\frac{l}{2}} \cdot \frac{\sigma_{BN}^{\varepsilon}}{\sqrt{n}})$$

$$< (\mu_{BN}^{\varepsilon} + Z_{\frac{l}{2}} \cdot \frac{\sigma_{BN}^{\varepsilon}}{\sqrt{n}}) < (\mu_{PN}^{\varepsilon} - Z_{\frac{l}{2}} \cdot \frac{\sigma_{PN}^{\varepsilon}}{\sqrt{n}}) < (\mu_{PN}^{\varepsilon} + Z_{\frac{l}{2}} \cdot \frac{\sigma_{PN}^{\varepsilon}}{\sqrt{n}}). \quad (8)$$

Proposition 1. *Under the condition (8), the range of the threshold α: $\alpha \in [\alpha^{min}, \alpha^{max}]$, where,*

$$\alpha^{min} = \frac{[(\mu_{PN}^{\varepsilon} - Z_{\frac{l}{2}} \cdot \frac{\sigma_{PN}^{\varepsilon}}{\sqrt{n}}) - (\mu_{BN}^{\varepsilon} + Z_{\frac{l}{2}} \cdot \frac{\sigma_{BN}^{\varepsilon}}{\sqrt{n}})]}{[(\mu_{PN}^{\varepsilon} - Z_{\frac{l}{2}} \cdot \frac{\sigma_{PN}^{\varepsilon}}{\sqrt{n}}) - (\mu_{BN}^{\varepsilon} + Z_{\frac{l}{2}} \cdot \frac{\sigma_{BN}^{\varepsilon}}{\sqrt{n}})] + [(\mu_{BP}^{\varepsilon} - Z_{\frac{l}{2}} \cdot \frac{\sigma_{BP}^{\varepsilon}}{\sqrt{n}}) - (\mu_{PP}^{\varepsilon} + Z_{\frac{l}{2}} \cdot \frac{\sigma_{PP}^{\varepsilon}}{\sqrt{n}})]}$$

$$\alpha^{max} = \min(\frac{[(\mu_{PN}^{\varepsilon} + Z_{\frac{l}{2}} \cdot \frac{\sigma_{PN}^{\varepsilon}}{\sqrt{n}}) - (\mu_{BN}^{\varepsilon} - Z_{\frac{l}{2}} \cdot \frac{\sigma_{BN}^{\varepsilon}}{\sqrt{n}})]}{[(\mu_{PN}^{\varepsilon} + Z_{\frac{l}{2}} \cdot \frac{\sigma_{PN}^{\varepsilon}}{\sqrt{n}}) - (\mu_{BN}^{\varepsilon} - Z_{\frac{l}{2}} \cdot \frac{\sigma_{BN}^{\varepsilon}}{\sqrt{n}})] + [(\mu_{BP}^{\varepsilon} + Z_{\frac{l}{2}} \cdot \frac{\sigma_{BP}^{\varepsilon}}{\sqrt{n}}) - (\mu_{PP}^{\varepsilon} - Z_{\frac{l}{2}} \cdot \frac{\sigma_{PP}^{\varepsilon}}{\sqrt{n}})]}, 1).$$

Proof. According to SDTRS, the expression of α is: $\alpha = \frac{(\lambda_{PN}^{\varepsilon} - \lambda_{BN}^{\varepsilon})}{(\lambda_{PN}^{\varepsilon} - \lambda_{BN}^{\varepsilon}) + (\lambda_{BP}^{\varepsilon} - \lambda_{PP}^{\varepsilon})}$.

Based on (8), we can get the relationships: $(\mu_{PN}^{\varepsilon} - Z_{\frac{l}{2}} \cdot \frac{\sigma_{PN}^{\varepsilon}}{\sqrt{n}}) - (\mu_{BN}^{\varepsilon} + Z_{\frac{l}{2}} \cdot \frac{\sigma_{BN}^{\varepsilon}}{\sqrt{n}}) \leq$

$\lambda_{PN}^{\varepsilon} - \lambda_{BN}^{\varepsilon} \leq \mu_{PN}^{\varepsilon} + Z_{\frac{l}{2}} \cdot \frac{\sigma_{PN}^{\varepsilon}}{\sqrt{n}}) - (\mu_{BN}^{\varepsilon} - Z_{\frac{l}{2}} \cdot \frac{\sigma_{BN}^{\varepsilon}}{\sqrt{n}})$, $(\mu_{BP}^{\varepsilon} - Z_{\frac{l}{2}} \cdot \frac{\sigma_{BP}^{\varepsilon}}{\sqrt{n}}) - (\mu_{PP}^{\varepsilon} +$

$Z_{\frac{l}{2}} \cdot \frac{\sigma_{PP}^{\varepsilon}}{\sqrt{n}}) \leq \lambda_{BP}^{\varepsilon} - \lambda_{PP}^{\varepsilon} \leq (\mu_{BP}^{\varepsilon} + Z_{\frac{l}{2}} \cdot \frac{\sigma_{BP}^{\varepsilon}}{\sqrt{n}}) - (\mu_{PP}^{\varepsilon} - Z_{\frac{l}{2}} \cdot \frac{\sigma_{PP}^{\varepsilon}}{\sqrt{n}})$. Therefore,

$[(\mu_{PN}^{\varepsilon} - Z_{\frac{l}{2}} \cdot \frac{\sigma_{PN}^{\varepsilon}}{\sqrt{n}}) - (\mu_{BN}^{\varepsilon} + Z_{\frac{l}{2}} \cdot \frac{\sigma_{BN}^{\varepsilon}}{\sqrt{n}})] + [(\mu_{BP}^{\varepsilon} - Z_{\frac{l}{2}} \cdot \frac{\sigma_{BP}^{\varepsilon}}{\sqrt{n}}) - (\mu_{PP}^{\varepsilon} + Z_{\frac{l}{2}} \cdot \frac{\sigma_{PP}^{\varepsilon}}{\sqrt{n}})] \leq$

$(\lambda_{PN}^{\varepsilon} - \lambda_{BN}^{\varepsilon}) + (\lambda_{BP}^{\varepsilon} - \lambda_{PP}^{\varepsilon}) \leq [(\mu_{PN}^{\varepsilon} + Z_{\frac{l}{2}} \cdot \frac{\sigma_{PN}^{\varepsilon}}{\sqrt{n}}) - (\mu_{BN}^{\varepsilon} - Z_{\frac{l}{2}} \cdot \frac{\sigma_{BN}^{\varepsilon}}{\sqrt{n}})] +$

$[(\mu_{BP}^{\varepsilon} + Z_{\frac{l}{2}} \cdot \frac{\sigma_{BP}^{\varepsilon}}{\sqrt{n}}) - (\mu_{PP}^{\varepsilon} - Z_{\frac{l}{2}} \cdot \frac{\sigma_{PP}^{\varepsilon}}{\sqrt{n}})]$. Furthermore, due to $\alpha \in [0, 1]$, we have:

$\alpha^{min} \leq \alpha = \frac{(\lambda_{PN}^{\varepsilon} - \lambda_{BN}^{\varepsilon})}{(\lambda_{PN}^{\varepsilon} - \lambda_{BN}^{\varepsilon}) + (\lambda_{BP}^{\varepsilon} - \lambda_{PP}^{\varepsilon})} \leq \alpha^{max}$. $\square$

Proposition 2. *Under the condition (8), the range of the threshold β: $\beta \in [\beta^{min}, \beta^{max}]$, where,*

$$\beta^{min} = \frac{[(\mu_{BN}^{\varepsilon} - Z_{\frac{l}{2}} \cdot \frac{\sigma_{BN}^{\varepsilon}}{\sqrt{n}}) - (\mu_{NN}^{\varepsilon} + Z_{\frac{l}{2}} \cdot \frac{\sigma_{NN}^{\varepsilon}}{\sqrt{n}})]}{[(\mu_{BN}^{\varepsilon} - Z_{\frac{l}{2}} \cdot \frac{\sigma_{BN}^{\varepsilon}}{\sqrt{n}}) - (\mu_{NN}^{\varepsilon} + Z_{\frac{l}{2}} \cdot \frac{\sigma_{NN}^{\varepsilon}}{\sqrt{n}})] + [(\mu_{NP}^{\varepsilon} - Z_{\frac{l}{2}} \cdot \frac{\sigma_{NP}^{\varepsilon}}{\sqrt{n}}) - (\mu_{BP}^{\varepsilon} + Z_{\frac{l}{2}} \cdot \frac{\sigma_{BP}^{\varepsilon}}{\sqrt{n}})]}$$

$$\beta^{max} = min(\frac{[(\mu_{BN}^{\varepsilon} + Z_{\frac{l}{2}} \cdot \frac{\sigma_{BN}^{\varepsilon}}{\sqrt{n}}) - (\mu_{NN}^{\varepsilon} - Z_{\frac{l}{2}} \cdot \frac{\sigma_{NN}^{\varepsilon}}{\sqrt{n}})]}{[(\mu_{BN}^{\varepsilon} + Z_{\frac{l}{2}} \cdot \frac{\sigma_{BN}^{\varepsilon}}{\sqrt{n}}) - (\mu_{NN}^{\varepsilon} - Z_{\frac{l}{2}} \cdot \frac{\sigma_{NN}^{\varepsilon}}{\sqrt{n}})] + [(\mu_{NP}^{\varepsilon} + Z_{\frac{l}{2}} \cdot \frac{\sigma_{NP}^{\varepsilon}}{\sqrt{n}}) - (\mu_{BP}^{\varepsilon} - Z_{\frac{l}{2}} \cdot \frac{\sigma_{BP}^{\varepsilon}}{\sqrt{n}})]}, 1).$$

Proof. According to SDTRS, the expression of β is: $\beta = \frac{(\lambda_{BN}^{\varepsilon} - \lambda_{NN}^{\varepsilon})}{(\lambda_{BN}^{\varepsilon} - \lambda_{NN}^{\varepsilon}) + (\lambda_{NP}^{\varepsilon} - \lambda_{BP}^{\varepsilon})}$.

Based on (8), we can get the relationships: $(\mu_{BN}^{\varepsilon} - Z_{\frac{l}{2}} \cdot \frac{\sigma_{BN}^{\varepsilon}}{\sqrt{n}}) - (\mu_{NN}^{\varepsilon} + Z_{\frac{l}{2}} \cdot \frac{\sigma_{NN}^{\varepsilon}}{\sqrt{n}}) \leq$

$\lambda_{BN}^{\varepsilon} - \lambda_{NN}^{\varepsilon} \leq \mu_{BN}^{\varepsilon} + Z_{\frac{l}{2}} \cdot \frac{\sigma_{BN}^{\varepsilon}}{\sqrt{n}}) - (\mu_{NN}^{\varepsilon} - Z_{\frac{l}{2}} \cdot \frac{\sigma_{NN}^{\varepsilon}}{\sqrt{n}})$, $(\mu_{NP}^{\varepsilon} - Z_{\frac{l}{2}} \cdot \frac{\sigma_{NP}^{\varepsilon}}{\sqrt{n}}) - (\mu_{BP}^{\varepsilon} +$

$Z_{\frac{l}{2}} \cdot \frac{\sigma_{BP}^{\varepsilon}}{\sqrt{n}}) \leq \lambda_{NP}^{\varepsilon} - \lambda_{BP}^{\varepsilon} \leq (\mu_{NP}^{\varepsilon} + Z_{\frac{l}{2}} \cdot \frac{\sigma_{NP}^{\varepsilon}}{\sqrt{n}}) - (\mu_{BP}^{\varepsilon} - Z_{\frac{l}{2}} \cdot \frac{\sigma_{BP}^{\varepsilon}}{\sqrt{n}})$. Therefore,

$[(\mu_{BN}^{\varepsilon} - Z_{\frac{l}{2}} \cdot \frac{\sigma_{BN}^{\varepsilon}}{\sqrt{n}}) - (\mu_{NN}^{\varepsilon} + Z_{\frac{l}{2}} \cdot \frac{\sigma_{NN}^{\varepsilon}}{\sqrt{n}})] + [(\mu_{NP}^{\varepsilon} - Z_{\frac{l}{2}} \cdot \frac{\sigma_{NP}^{\varepsilon}}{\sqrt{n}}) - (\mu_{BP}^{\varepsilon} + Z_{\frac{l}{2}} \cdot \frac{\sigma_{BP}^{\varepsilon}}{\sqrt{n}})] \leq$

$(\lambda_{BN}^{\varepsilon} - \lambda_{NN}^{\varepsilon}) + (\lambda_{NP}^{\varepsilon} - \lambda_{BP}^{\varepsilon}) \leq [(\mu_{BN}^{\varepsilon} + Z_{\frac{l}{2}} \cdot \frac{\sigma_{BN}^{\varepsilon}}{\sqrt{n}}) - (\mu_{NN}^{\varepsilon} - Z_{\frac{l}{2}} \cdot \frac{\sigma_{NN}^{\varepsilon}}{\sqrt{n}})] +$

$[(\mu_{NP}^{\varepsilon} + Z_{\frac{l}{2}} \cdot \frac{\sigma_{NP}^{\varepsilon}}{\sqrt{n}}) - (\mu_{BP}^{\varepsilon} - Z_{\frac{l}{2}} \cdot \frac{\sigma_{BP}^{\varepsilon}}{\sqrt{n}})]$. Furthermore, due to $\beta \in [0, 1]$, we have:

$\beta^{min} \leq \beta = \frac{(\lambda_{BN}^{\varepsilon} - \lambda_{NN}^{\varepsilon})}{(\lambda_{BN}^{\varepsilon} - \lambda_{NN}^{\varepsilon}) + (\lambda_{NP}^{\varepsilon} - \lambda_{BP}^{\varepsilon})} \leq \beta^{max}$. $\square$

Obviously, $(\alpha_L, \alpha_U) \subseteq [\alpha^{min}, \alpha^{max}]$, $(\beta_L, \beta_U) \subseteq [\beta^{min}, \beta^{max}]$. Note that, in this model, the width of $[\mu_{\bullet\bullet}^{\varepsilon} - Z_{\frac{l}{2}} \cdot \frac{\sigma_{\bullet\bullet}^{\varepsilon}}{\sqrt{n}}, \mu_{\bullet\bullet}^{\varepsilon} + Z_{\frac{l}{2}} \cdot \frac{\sigma_{\bullet\bullet}^{\varepsilon}}{\sqrt{n}}]$ is depended on the significance level l and the sample size when $\mu_{\bullet\bullet}^{\varepsilon}, \sigma_{\bullet\bullet}^{\varepsilon}$ is known. The width of the interval, *i.e.*, the margin of error, gets smaller as the sample size increases. In the same way, the width of the interval increases when the confidence decreases.

3.1.2 Confidence Intervals of $\lambda_{\bullet\bullet}^{\varepsilon}$ (σ Unknown)

Suppose $\lambda_{\bullet\bullet}^{\varepsilon} = \{\lambda_{\bullet\bullet}^{\varepsilon 1}, \lambda_{\bullet\bullet}^{\varepsilon 2}, \cdots, \lambda_{\bullet\bullet}^{\varepsilon n}\}$ are the n observed random samples. We denote the sample mean $\overline{\lambda}_{\bullet\bullet}^{\varepsilon} = \frac{1}{n} \cdot \sum_{i=1}^{n} \lambda_{\bullet\bullet}^{\varepsilon i}$, and the sample standard deviation $S_{\bullet\bullet}^{\varepsilon} = \sqrt{\frac{\sum_{i=1}^{n}(\lambda_{\bullet\bullet}^{\varepsilon i} - \overline{\lambda}_{\bullet\bullet}^{\varepsilon})^2}{n-1}}$. If the population of $\lambda_{\bullet\bullet}^{\varepsilon}$ is normally distributed or $n \geq 30$. Given a significance level l, the estimation of $\lambda_{\bullet\bullet}^{\varepsilon}$ can be calculated as:

$$\overline{\lambda}_{\bullet\bullet}^{\varepsilon} - t_{\frac{l}{2},n-1} \cdot \frac{S_{\bullet\bullet}^{\varepsilon}}{\sqrt{n}} \leq \lambda_{\bullet\bullet}^{\varepsilon} \leq \overline{\lambda}_{\bullet\bullet}^{\varepsilon} + t_{\frac{l}{2},n-1} \cdot \frac{S_{\bullet\bullet}^{\varepsilon}}{\sqrt{n}}. \tag{9}$$

Specially, suppose the samples are relative to population with $n/N > 0.05$ (n denotes the size of observed samples and N denotes the size of population samples). We use the finite population correction factor to correct the estimations, and (9) can be rewritten as:

$$\overline{\lambda}_{\bullet\bullet}^{\varepsilon} - t_{\frac{\alpha}{2},n-1} \cdot \frac{S_{\bullet\bullet}^{\varepsilon}}{\sqrt{n}} \cdot \sqrt{\frac{N-n}{N-1}} \leq \lambda_{\bullet\bullet}^{\varepsilon} \leq \overline{\lambda}_{\bullet\bullet}^{\varepsilon} + t_{\frac{\alpha}{2},n-1} \cdot \frac{S_{\bullet\bullet}^{\varepsilon}}{\sqrt{n}} \cdot \sqrt{\frac{N-n}{N-1}}. \tag{10}$$

In (9) and (10), there are two steps to find the required t critical value: (i) Determine the degrees of freedom: $df = (n-1)$; (ii) Use the appropriate confidence level and the df and locate the t critical value in the t critical value table. For simplicity, we only consider (9) in our following studies. As we stated in Sect. 3.1.1, one can use an interval $[\overline{\lambda}_{\bullet\bullet}^{\varepsilon} - t_{\frac{l}{2},n-1} \cdot \frac{S_{\bullet\bullet}^{\varepsilon}}{\sqrt{n}}, \overline{\lambda}_{\bullet\bullet}^{\varepsilon} + t_{\frac{l}{2},n-1} \cdot \frac{S_{\bullet\bullet}^{\varepsilon}}{\sqrt{n}}]$ to estimate $\lambda_{\bullet\bullet}^{\varepsilon}$ by choosing a proper l, and the 3×2 matrix can be expressed as:

	$C\ (P)$	$\neg C\ (N)$
a_P	$[\overline{\lambda}_{PP}^{\varepsilon}-t_{\frac{l}{2},n-1}\cdot\frac{S_{PP}^{\varepsilon}}{\sqrt{n}},\overline{\lambda}_{PP}^{\varepsilon}+t_{\frac{l}{2},n-1}\cdot\frac{S_{PP}^{\varepsilon}}{\sqrt{n}}]$	$[\overline{\lambda}_{PN}^{\varepsilon}-t_{\frac{l}{2},n-1}\cdot\frac{S_{PN}^{\varepsilon}}{\sqrt{n}},\overline{\lambda}_{PN}^{\varepsilon}+t_{\frac{l}{2},n-1}\cdot\frac{S_{PN}^{\varepsilon}}{\sqrt{n}}]$
a_B	$[\overline{\lambda}_{BP}^{\varepsilon}-t_{\frac{l}{2},n-1}\cdot\frac{S_{BP}^{\varepsilon}}{\sqrt{n}},\overline{\lambda}_{BP}^{\varepsilon}+t_{\frac{l}{2},n-1}\cdot\frac{S_{BP}^{\varepsilon}}{\sqrt{n}}]$	$[\overline{\lambda}_{BN}^{\varepsilon}-t_{\frac{l}{2},n-1}\cdot\frac{S_{BN}^{\varepsilon}}{\sqrt{n}},\overline{\lambda}_{BN}^{\varepsilon}+t_{\frac{l}{2},n-1}\cdot\frac{S_{BN}^{\varepsilon}}{\sqrt{n}}]$
a_N	$[\overline{\lambda}_{NP}^{\varepsilon}-t_{\frac{l}{2},n-1}\cdot\frac{S_{NP}^{\varepsilon}}{\sqrt{n}},\overline{\lambda}_{NP}^{\varepsilon}+t_{\frac{l}{2},n-1}\cdot\frac{S_{NP}^{\varepsilon}}{\sqrt{n}}]$	$[\overline{\lambda}_{NN}^{\varepsilon}-t_{\frac{l}{2},n-1}\cdot\frac{S_{NN}^{\varepsilon}}{\sqrt{n}},\overline{\lambda}_{NN}^{\varepsilon}+t_{\frac{l}{2},n-1}\cdot\frac{S_{NN}^{\varepsilon}}{\sqrt{n}}]$

Analogously, following our discussions in Sect. 3.3.1, we also consider two case: the lower bound and the upper bound of $\lambda_{\bullet\bullet}^{\varepsilon}$. On the basis of conditions reported in (4), the losses in this model satisfy the following constraints:

$$(\overline{\lambda}_{PP}^{\varepsilon} - t_{\frac{l}{2},n-1} \cdot \frac{S_{PP}^{\varepsilon}}{\sqrt{n}}) \leq (\overline{\lambda}_{BP}^{\varepsilon} - t_{\frac{l}{2},n-1} \cdot \frac{S_{BP}^{\varepsilon}}{\sqrt{n}}) \leq (\overline{\lambda}_{NP}^{\varepsilon} - t_{\frac{l}{2},n-1} \cdot \frac{S_{NP}^{\varepsilon}}{\sqrt{n}});$$

$$(\overline{\lambda}_{NN}^{\varepsilon} - t_{\frac{l}{2},n-1} \cdot \frac{S_{NN}^{\varepsilon}}{\sqrt{n}}) \leq (\overline{\lambda}_{BN}^{\varepsilon} - t_{\frac{l}{2},n-1} \cdot \frac{S_{BN}^{\varepsilon}}{\sqrt{n}}) \leq (\overline{\lambda}_{PN}^{\varepsilon} - t_{\frac{l}{2},n-1} \cdot \frac{S_{PN}^{\varepsilon}}{\sqrt{n}});$$

$$(\overline{\lambda}_{PP}^{\varepsilon} + t_{\frac{l}{2},n-1} \cdot \frac{S_{PP}^{\varepsilon}}{\sqrt{n}}) \leq (\overline{\lambda}_{BP}^{\varepsilon} + t_{\frac{l}{2},n-1} \cdot \frac{S_{BP}^{\varepsilon}}{\sqrt{n}}) \leq (\overline{\lambda}_{NP}^{\varepsilon} + t_{\frac{l}{2},n-1} \cdot \frac{S_{NP}^{\varepsilon}}{\sqrt{n}});$$

$$(\overline{\lambda}_{NN}^{\varepsilon} + t_{\frac{l}{2},n-1} \cdot \frac{S_{NN}^{\varepsilon}}{\sqrt{n}}) \leq (\overline{\lambda}_{BN}^{\varepsilon} + t_{\frac{l}{2},n-1} \cdot \frac{S_{BN}^{\varepsilon}}{\sqrt{n}}) \leq (\overline{\lambda}_{PN}^{\varepsilon} + t_{\frac{l}{2},n-1} \cdot \frac{S_{PN}^{\varepsilon}}{\sqrt{n}}).$$

$$\tag{11}$$

Since $Pr(C|[x]) + Pr(\neg C|[x]) = 1$, the threshold values under the lower and upper bounds are calculated.

$$\alpha_L = \frac{[(\overline{\lambda}^\varepsilon_{PN} - t_{\frac{l}{2},n-1} \cdot \frac{S^\varepsilon_{PN}}{\sqrt{n}}) - (\overline{\lambda}^\varepsilon_{BN} - t_{\frac{l}{2},n-1} \cdot \frac{S^\varepsilon_{BN}}{\sqrt{n}})]}{[(\overline{\lambda}^\varepsilon_{PN} - t_{\frac{l}{2},n-1} \cdot \frac{S^\varepsilon_{PN}}{\sqrt{n}}) - (\overline{\lambda}^\varepsilon_{BN} - t_{\frac{l}{2},n-1} \cdot \frac{S^\varepsilon_{BN}}{\sqrt{n}})] + [(\overline{\lambda}^\varepsilon_{BP} - t_{\frac{l}{2},n-1} \cdot \frac{S^\varepsilon_{BP}}{\sqrt{n}}) - (\overline{\lambda}^\varepsilon_{PP} - t_{\frac{l}{2},n-1} \cdot \frac{S^\varepsilon_{PP}}{\sqrt{n}})]},$$

$$\beta_L = \frac{[(\overline{\lambda}^\varepsilon_{BN} - t_{\frac{l}{2},n-1} \cdot \frac{S^\varepsilon_{BN}}{\sqrt{n}}) - (\overline{\lambda}^\varepsilon_{NN} - t_{\frac{l}{2},n-1} \cdot \frac{S^\varepsilon_{NN}}{\sqrt{n}})]}{[(\overline{\lambda}^\varepsilon_{BN} - t_{\frac{l}{2},n-1} \cdot \frac{S^\varepsilon_{BN}}{\sqrt{n}}) - (\overline{\lambda}^\varepsilon_{NN} - t_{\frac{l}{2},n-1} \cdot \frac{S^\varepsilon_{NN}}{\sqrt{n}})] + [(\overline{\lambda}^\varepsilon_{NP} - t_{\frac{l}{2},n-1} \cdot \frac{S^\varepsilon_{NP}}{\sqrt{n}}) - (\overline{\lambda}^\varepsilon_{BP} - t_{\frac{l}{2},n-1} \cdot \frac{S^\varepsilon_{BP}}{\sqrt{n}})]},$$

$$\gamma_L = \frac{[(\overline{\lambda}^\varepsilon_{PN} - t_{\frac{l}{2},n-1} \cdot \frac{S^\varepsilon_{PN}}{\sqrt{n}}) - (\overline{\lambda}^\varepsilon_{NN} - t_{\frac{l}{2},n-1} \cdot \frac{S^\varepsilon_{NN}}{\sqrt{n}})]}{[(\overline{\lambda}^\varepsilon_{PN} - t_{\frac{l}{2},n-1} \cdot \frac{S^\varepsilon_{PN}}{\sqrt{n}}) - (\overline{\lambda}^\varepsilon_{NN} - t_{\frac{l}{2},n-1} \cdot \frac{S^\varepsilon_{NN}}{\sqrt{n}})] + [(\overline{\lambda}^\varepsilon_{NP} - t_{\frac{l}{2},n-1} \cdot \frac{S^\varepsilon_{NP}}{\sqrt{n}}) - (\overline{\lambda}^\varepsilon_{PP} - t_{\frac{l}{2},n-1} \cdot \frac{S^\varepsilon_{PP}}{\sqrt{n}})]},$$

$$\alpha_U = \frac{[(\overline{\lambda}^\varepsilon_{PN} + t_{\frac{l}{2},n-1} \cdot \frac{S^\varepsilon_{PN}}{\sqrt{n}}) - (\overline{\lambda}^\varepsilon_{BN} + t_{\frac{l}{2},n-1} \cdot \frac{S^\varepsilon_{BN}}{\sqrt{n}})]}{[(\overline{\lambda}^\varepsilon_{PN} + t_{\frac{l}{2},n-1} \cdot \frac{S^\varepsilon_{PN}}{\sqrt{n}}) - (\overline{\lambda}^\varepsilon_{BN} + t_{\frac{l}{2},n-1} \cdot \frac{S^\varepsilon_{BN}}{\sqrt{n}})] + [(\overline{\lambda}^\varepsilon_{BP} + t_{\frac{l}{2},n-1} \cdot \frac{S^\varepsilon_{BP}}{\sqrt{n}}) - (\overline{\lambda}^\varepsilon_{PP} + t_{\frac{l}{2},n-1} \cdot \frac{S^\varepsilon_{PP}}{\sqrt{n}})]},$$

$$\beta_U = \frac{[(\overline{\lambda}^\varepsilon_{BN} + t_{\frac{l}{2},n-1} \cdot \frac{S^\varepsilon_{BN}}{\sqrt{n}}) - (\overline{\lambda}^\varepsilon_{NN} + t_{\frac{l}{2},n-1} \cdot \frac{S^\varepsilon_{NN}}{\sqrt{n}})]}{[(\overline{\lambda}^\varepsilon_{BN} + t_{\frac{l}{2},n-1} \cdot \frac{S^\varepsilon_{BN}}{\sqrt{n}}) - (\overline{\lambda}^\varepsilon_{NN} + t_{\frac{l}{2},n-1} \cdot \frac{S^\varepsilon_{NN}}{\sqrt{n}})] + [(\overline{\lambda}^\varepsilon_{NP} + t_{\frac{l}{2},n-1} \cdot \frac{S^\varepsilon_{NP}}{\sqrt{n}}) - (\overline{\lambda}^\varepsilon_{BP} + t_{\frac{l}{2},n-1} \cdot \frac{S^\varepsilon_{BP}}{\sqrt{n}})]},$$

$$\gamma_U = \frac{[(\overline{\lambda}^\varepsilon_{PN} + t_{\frac{l}{2},n-1} \cdot \frac{S^\varepsilon_{PN}}{\sqrt{n}}) - (\overline{\lambda}^\varepsilon_{NN} + t_{\frac{l}{2},n-1} \cdot \frac{S^\varepsilon_{NN}}{\sqrt{n}})]}{[(\overline{\lambda}^\varepsilon_{PN} + t_{\frac{l}{2},n-1} \cdot \frac{S^\varepsilon_{PN}}{\sqrt{n}}) - (\overline{\lambda}^\varepsilon_{NN} + t_{\frac{l}{2},n-1} \cdot \frac{S^\varepsilon_{NN}}{\sqrt{n}})] + [(\overline{\lambda}^\varepsilon_{NP} + t_{\frac{l}{2},n-1} \cdot \frac{S^\varepsilon_{NP}}{\sqrt{n}}) - (\overline{\lambda}^\varepsilon_{PP} + t_{\frac{l}{2},n-1} \cdot \frac{S^\varepsilon_{PP}}{\sqrt{n}})]}.$$

Furthermore, we consider one type of stricter condition based on (11) as follows.

$$(\overline{\lambda}^\varepsilon_{PP} - t_{\frac{1}{2},n-1} \cdot \frac{S^\varepsilon_{PP}}{\sqrt{n}}) < (\overline{\lambda}^\varepsilon_{PP} + t_{\frac{1}{2},n-1} \cdot \frac{S^\varepsilon_{PP}}{\sqrt{n}}) \le (\overline{\lambda}^\varepsilon_{BP} - t_{\frac{1}{2},n-1} \cdot \frac{S^\varepsilon_{BP}}{\sqrt{n}})$$

$$< (\overline{\lambda}^\varepsilon_{BP} + t_{\frac{1}{2},n-1} \cdot \frac{S^\varepsilon_{BP}}{\sqrt{n}}) < (\overline{\lambda}^\varepsilon_{NP} - t_{\frac{1}{2},n-1} \cdot \frac{S^\varepsilon_{NP}}{\sqrt{n}}) < (\overline{\lambda}^\varepsilon_{NP} + t_{\frac{1}{2},n-1} \cdot \frac{S^\varepsilon_{NP}}{\sqrt{n}});$$

$$(\overline{\lambda}^\varepsilon_{NN} - t_{\frac{1}{2},n-1} \cdot \frac{S^\varepsilon_{NN}}{\sqrt{n}}) < (\overline{\lambda}^\varepsilon_{NN} + t_{\frac{1}{2},n-1} \cdot \frac{S^\varepsilon_{NN}}{\sqrt{n}}) \le (\overline{\lambda}^\varepsilon_{BN} - t_{\frac{1}{2},n-1} \cdot \frac{S^\varepsilon_{BN}}{\sqrt{n}})$$

$$< (\overline{\lambda}^\varepsilon_{BN} + t_{\frac{1}{2},n-1} \cdot \frac{S^\varepsilon_{BN}}{\sqrt{n}}) < (\overline{\lambda}^\varepsilon_{PN} - t_{\frac{1}{2},n-1} \cdot \frac{S^\varepsilon_{PN}}{\sqrt{n}}) < (\overline{\lambda}^\varepsilon_{PN} + t_{\frac{1}{2},n-1} \cdot \frac{S^\varepsilon_{PN}}{\sqrt{n}}). \tag{12}$$

Proposition 3. *Under the condition (12), the range of the threshold α: $\alpha \in [\alpha^{min}, \alpha^{max}]$, where,*

$$\alpha^{min} = \frac{[(\overline{\lambda}^\varepsilon_{PN} - t_{\frac{l}{2},n-1} \cdot \frac{S^\varepsilon_{PN}}{\sqrt{n}}) - (\overline{\lambda}^\varepsilon_{BN} + t_{\frac{l}{2},n-1} \cdot \frac{S^\varepsilon_{BN}}{\sqrt{n}})]}{[(\overline{\lambda}^\varepsilon_{PN} - t_{\frac{l}{2},n-1} \cdot \frac{S^\varepsilon_{PN}}{\sqrt{n}}) - (\overline{\lambda}^\varepsilon_{BN} + t_{\frac{l}{2},n-1} \cdot \frac{S^\varepsilon_{BN}}{\sqrt{n}})] + [(\overline{\lambda}^\varepsilon_{BP} - t_{\frac{l}{2},n-1} \cdot \frac{S^\varepsilon_{BP}}{\sqrt{n}}) - (\overline{\lambda}^\varepsilon_{PP} + t_{\frac{l}{2},n-1} \cdot \frac{S^\varepsilon_{PP}}{\sqrt{n}})]}$$

$$\alpha^{max} = min(\frac{[(\overline{\lambda}^\varepsilon_{PN} + t_{\frac{l}{2},n-1} \cdot \frac{S^\varepsilon_{PN}}{\sqrt{n}}) - (\overline{\lambda}^\varepsilon_{BN} - t_{\frac{l}{2},n-1} \cdot \frac{S^\varepsilon_{BN}}{\sqrt{n}})]}{[(\overline{\lambda}^\varepsilon_{PN} + t_{\frac{l}{2},n-1} \cdot \frac{S^\varepsilon_{PN}}{\sqrt{n}}) - (\overline{\lambda}^\varepsilon_{BN} - t_{\frac{l}{2},n-1} \cdot \frac{S^\varepsilon_{BN}}{\sqrt{n}})] + [(\overline{\lambda}^\varepsilon_{BP} + t_{\frac{l}{2},n-1} \cdot \frac{S^\varepsilon_{BP}}{\sqrt{n}}) - (\overline{\lambda}^\varepsilon_{PP} - t_{\frac{l}{2},n-1} \cdot \frac{S^\varepsilon_{PP}}{\sqrt{n}})]}, 1).$$

Proof. It is similar to the proof of Proposition 1. □

Proposition 4. *Under the condition (12), the range of the threshold β: $\beta \in [\beta^{min}, \beta^{max}]$, where,*

$$\beta^{min} = \frac{[(\overline{\lambda}^\varepsilon_{BN} - t_{\frac{1}{2},n-1} \cdot \frac{S^\varepsilon_{BN}}{\sqrt{n}}) - (\overline{\lambda}^\varepsilon_{NN} + t_{\frac{1}{2},n-1} \cdot \frac{S^\varepsilon_{NN}}{\sqrt{n}})]}{[(\overline{\lambda}^\varepsilon_{BN} - t_{\frac{1}{2},n-1} \cdot \frac{S^\varepsilon_{BN}}{\sqrt{n}}) - (\overline{\lambda}^\varepsilon_{NN} + t_{\frac{1}{2},n-1} \cdot \frac{S^\varepsilon_{NN}}{\sqrt{n}})] + [(\overline{\lambda}^\varepsilon_{NP} - t_{\frac{1}{2},n-1} \cdot \frac{S^\varepsilon_{NP}}{\sqrt{n}}) - (\overline{\lambda}^\varepsilon_{BP} + t_{\frac{1}{2},n-1} \cdot \frac{S^\varepsilon_{BP}}{\sqrt{n}})]},$$

$$\beta^{max} = min(\frac{[(\overline{\lambda}^\varepsilon_{BN} + t_{\frac{1}{2},n-1} \cdot \frac{S^\varepsilon_{BN}}{\sqrt{n}}) - (\overline{\lambda}^\varepsilon_{NN} - t_{\frac{1}{2},n-1} \cdot \frac{S^\varepsilon_{NN}}{\sqrt{n}})]}{[(\overline{\lambda}^\varepsilon_{BN} + t_{\frac{1}{2},n-1} \cdot \frac{S^\varepsilon_{BN}}{\sqrt{n}}) - (\overline{\lambda}^\varepsilon_{NN} - t_{\frac{1}{2},n-1} \cdot \frac{S^\varepsilon_{NN}}{\sqrt{n}})] + [(\overline{\lambda}^\varepsilon_{NP} + t_{\frac{1}{2},n-1} \cdot \frac{S^\varepsilon_{NP}}{\sqrt{n}}) - (\overline{\lambda}^\varepsilon_{BP} - t_{\frac{1}{2},n-1} \cdot \frac{S^\varepsilon_{BP}}{\sqrt{n}})]}, 1).$$

Proof. It is similar to the proof of Proposition 2.   □

Obviously, $(\alpha_L, \alpha_U) \subseteq [\alpha^{min}, \alpha^{max}]$, $(\beta_L, \beta_U) \subseteq [\beta^{min}, \beta^{max}]$.

3.2 Stochastic Decision-Theoretic Rough Sets with Statistical Distributions

Following our discussions in Sects. 2.2 and 3.1, in this subsection, we mainly discuss the two scenarios that the loss functions obey uniform distribution $\lambda^\varepsilon_{\bullet\bullet} \sim U(a, b)$ and normal distribution $\lambda^\varepsilon_{\bullet\bullet} \sim N(\mu, \sigma^2)$, respectively.

3.2.1 SDTRS with Uniform Distribution

In statistics, the uniform distribution is a family of probability distributions such that for each member of the family, all intervals of the same length on the distribution's support are equally probable. The support is defined by the two parameters, a and b, which are its minimum and maximum values. The distribution is abbreviated $U(a, b)$. The probability density function of the uniform distribution is:

$$f(x) = \begin{cases} \frac{1}{b-a} & a \leq x \leq b \\ 0 & \text{others} \end{cases}$$

In our discussions, suppose $\lambda^\varepsilon_{PP} \sim U(a_{PP}, b_{PP})$, $\lambda^\varepsilon_{BP} \sim U(a_{BP}, b_{BP})$, $\lambda^\varepsilon_{NP} \sim U(a_{NP}, b_{NP})$; $\lambda^\varepsilon_{NN} \sim U(a_{NN}, b_{NN})$, $\lambda^\varepsilon_{BN} \sim U(a_{BN}, b_{BN})$, $\lambda^\varepsilon_{PN} \sim U(a_{PN}, b_{PN})$. According to Sect. 2.2, we choose the expected values of $\overline{\lambda}^\varepsilon_{\bullet\bullet}$ ($\bullet = P, B, N$) as the measure to estimate $\lambda^\varepsilon_{\bullet\bullet}$. Considered the conditions of $a_{PP} \leq a_{BP} < a_{NP}$, $a_{NN} \leq a_{BN} < a_{PN}$; $b_{PP} \leq b_{BP} < b_{NP}$, $b_{NN} \leq b_{BN} < b_{PN}$, we have $\frac{a_{PP}+b_{PP}}{2} \leq \frac{a_{BP}+b_{BP}}{2} < \frac{a_{NP}+b_{NP}}{2}$ and $\frac{a_{NN}+b_{NN}}{2} \leq \frac{a_{BN}+b_{BN}}{2} < \frac{a_{PN}+b_{PN}}{2}$, that is, $\overline{\lambda}^\varepsilon_{PP} \leq \overline{\lambda}^\varepsilon_{BP} < \overline{\lambda}^\varepsilon_{NP}$, $\overline{\lambda}^\varepsilon_{NN} \leq \overline{\lambda}^\varepsilon_{BN} < \overline{\lambda}^\varepsilon_{PN}$. Therefore, we directly utilize $\overline{\lambda}^\varepsilon_{\bullet\bullet}$ to replace $\lambda^\varepsilon_{\bullet\bullet}$ and get the $\overline{\alpha}, \overline{\beta}, \overline{\gamma}$ as:

$$\overline{\alpha} = \frac{\overline{\lambda}^\varepsilon_{PN} - \overline{\lambda}^\varepsilon_{BN}}{(\overline{\lambda}^\varepsilon_{PN} - \overline{\lambda}^\varepsilon_{BN}) + (\overline{\lambda}^\varepsilon_{BP}) - \overline{\lambda}^\varepsilon_{PP})}$$

$$= \frac{\frac{(a_{PN}+b_{PN})}{2} - \frac{(a_{BN}+b_{BN})}{2}}{[\frac{(a_{PN}+b_{PN})}{2} - \frac{(a_{BN}+b_{BN})}{2}] + [\frac{(a_{BP}+b_{BP})}{2} - \frac{(a_{PP}+b_{PP})}{2}]}$$

$$= \frac{(a_{PN} + b_{PN}) - (a_{BN} + b_{BN})}{[(a_{PN} + b_{PN}) - (a_{BN} + b_{BN})] + [(a_{BP} + b_{BP}) - (a_{PP} + b_{PP})]},$$

$$\overline{\beta} = \frac{\overline{\lambda}^\varepsilon_{BN} - \overline{\lambda}^\varepsilon_{NN}}{(\overline{\lambda}^\varepsilon_{BN} - \overline{\lambda}^\varepsilon_{NN}) + (\overline{\lambda}^\varepsilon_{NP} - \overline{\lambda}^\varepsilon_{BP})}$$

$$= \frac{\frac{(a_{BN}+b_{BN})}{2} - \frac{(a_{NN}+b_{NN})}{2}}{[\frac{(a_{BN}+b_{BN})}{2} - \frac{(a_{NN}+b_{NN})}{2}] + [\frac{(a_{NP}+b_{NP})}{2} - \frac{(a_{BP}+b_{BP})}{2}]}$$

$$= \frac{(a_{BN} + b_{BN}) - (a_{NN} + b_{NN})}{[(a_{BN} + b_{BN}) - (a_{NN} + b_{NN})] + [(a_{NP} + b_{NP}) - [a_{BP} + b_{BP}]]},$$

$$\overline{\gamma} = \frac{\overline{\lambda}^\varepsilon_{PN} - \overline{\lambda}^\varepsilon_{NN}}{(\overline{\lambda}^\varepsilon_{PN} - \overline{\lambda}^\varepsilon_{NN}) + (\overline{\lambda}^\varepsilon_{NP} - \overline{\lambda}^\varepsilon_{PP})}$$

$$= \frac{\frac{(a_{PN}+b_{PN})}{2} - \frac{(a_{NN}+b_{NN})}{2}}{[\frac{(a_{PN}+b_{PN})}{2} - \frac{(a_{NN}+b_{NN})}{2}] + [\frac{(a_{NP}+b_{NP})}{2} - \frac{(a_{PP}+b_{PP})}{2}]}$$

$$= \frac{(a_{PN} + b_{PN}) - (a_{NN} + b_{NN})}{[(a_{PN} + b_{PN}) - (a_{NN} + b_{NN})] + [(a_{NP} + b_{NP}) - (a_{PP} + b_{PP})]}.$$

3.2.2   SDTRS with Normal Distribution

The normal distribution is a continuous probability distribution that has a bell-shaped probability density function, known as the Gaussian function or informally the bell curve:

$$f(x; \mu, \sigma^2) = \frac{1}{\sigma\sqrt{2\pi}} e^{-\frac{(x-\mu)^2}{2\sigma^2}} \tag{13}$$

where parameter μ is the mean or expectation (location of the peak) and σ^2 is the variance. σ is known as the standard deviation. The distribution with $\mu = 0$ and $\sigma^2 = 1$ is called the standard normal. A normal distribution is often used as a first approximation to describe real-valued random variables that cluster around a single mean value.

The cumulative distribution function (CDF) of the standard normal distribution, usually denoted with, Φ is the integral

$$\Phi(x) = \frac{1}{\sqrt{2\pi}} \int_{-\infty}^{x} e^{-z^2/2} dz \tag{14}$$

For a generic normal distribution f with mean μ and deviation σ, the cumulative distribution function is

$$F(x) = \Phi(\frac{x - \mu}{\sigma}) = \frac{1}{2}[1 + erf(\frac{x - \mu}{\sigma\sqrt{2}})] \tag{15}$$

where, the error function $erf(x) = \frac{1}{\sqrt{\pi}} \int_{-x}^{x} e^{-z^2} dz$ describes the probability of a random variable with normal distribution of mean 0 and variance $\frac{1}{2}$ falling in the range.

The probability that a normal deviate lies in the range x_1 and x_2 $(x_2 < x_1)$ is given by:

$$Pr(x_2 \leq x \leq x_1) = F(x_1) - F(x_2) = \Phi(\frac{x_1 - \mu}{\sigma}) - \Phi(\frac{x_2 - \mu}{\sigma}) \tag{16}$$

For simplicity, we introduce "68-95-99.7" rule to develop the SDTRS model with normal distribution. The "68-95-99.7" rule, also known as the three-sigma rule or empirical rule, states that nearly all values lie within three standard deviations of the mean in a normal distribution. About 68 % of values drawn from a normal distribution are within one standard deviation σ away from the mean; about 95 % of the values lie within two standard deviations; and about 99.7 % are within three standard deviations. Here, we introduce confidence intervals to our study, and three confidence intervals of $\lambda_{\bullet\bullet}^{\varepsilon}$ ($\bullet = P, B, N$) can be expressed as:

$$Pr(\mu_{\bullet\bullet}^{\varepsilon} - \sigma_{\bullet\bullet}^{\varepsilon} \leq \lambda_{\bullet\bullet}^{\varepsilon} \leq \mu_{\bullet\bullet}^{\varepsilon} + \sigma_{\bullet\bullet}^{\varepsilon}) = \Phi(1) - \Phi(-1) = erf(\frac{1}{\sqrt{2}}) \approx 0.6827;$$

$$Pr(\mu_{\bullet\bullet}^{\varepsilon} - 2\sigma_{\bullet\bullet}^{\varepsilon} \leq \lambda_{\bullet\bullet}^{\varepsilon} \leq \mu_{\bullet\bullet}^{\varepsilon} + 2\sigma_{\bullet\bullet}^{\varepsilon}) = \Phi(2) - \Phi(-2) = erf(\frac{2}{\sqrt{2}}) \approx 0.9545;$$

$$Pr(\mu_{\bullet\bullet}^{\varepsilon} - 3\sigma_{\bullet\bullet}^{\varepsilon} \leq \lambda_{\bullet\bullet}^{\varepsilon} \leq \mu_{\bullet\bullet}^{\varepsilon} + 3\sigma_{\bullet\bullet}^{\varepsilon}) = \Phi(3) - \Phi(-3) = erf(\frac{3}{\sqrt{2}}) \approx 0.9973.$$

Suppose $\lambda_{PP}^{\varepsilon} \sim N(\mu_{PP}, \sigma_{PP}^2)$, $\lambda_{BP}^{\varepsilon} \sim N(\mu_{BP}, \sigma_{BP}^2)$, $\lambda_{NP}^{\varepsilon} \sim N(\mu_{NP}, \sigma_{NP}^2)$; $\lambda_{NN}^{\varepsilon} \sim N(\mu_{NN}, \sigma_{NN}^2)$, $\lambda_{BN}^{\varepsilon} \sim N(\mu_{BN}, \sigma_{BN}^2)$, $\lambda_{PN}^{\varepsilon} \sim N(\mu_{PN}, \sigma_{PN}^2)$. In our following discussions, we utilize the intervals $[\mu_{\bullet\bullet}^{\varepsilon} - n\sigma_{\bullet\bullet}^{\varepsilon}, \mu_{\bullet\bullet}^{\varepsilon} + n\sigma_{\bullet\bullet}^{\varepsilon}]$ ($n = 1, 2, 3$) instead of $\lambda_{\bullet\bullet}^{\varepsilon}$.

On the basis of conditions reported in (4), the losses in this model satisfy: $\mu_{\bullet\bullet}^{\varepsilon} + n\sigma_{\bullet\bullet}^{\varepsilon} > \mu_{\bullet\bullet}^{\varepsilon} - n\sigma_{\bullet\bullet}^{\varepsilon} \geq 0$. Specially, we consider two scenarios: (1) $\lambda_{\bullet\bullet}^{\varepsilon} = \mu_{\bullet\bullet}^{\varepsilon} - n\sigma_{\bullet\bullet}^{\varepsilon}$, (2) $\lambda_{\bullet\bullet}^{\varepsilon} = \mu_{\bullet\bullet}^{\varepsilon} + n\sigma_{\bullet\bullet}^{\varepsilon}$. The threshold values under above two cases $(\alpha^L, \beta^L, \gamma^L)$ and $(\alpha^U, \beta^U, \gamma^U)$ can be calculated as follows:

$$\alpha^L = \frac{(\mu_{PN}^{\varepsilon} - n\sigma_{PN}^{\varepsilon}) - (\mu_{BN}^{\varepsilon} - n\sigma_{BN}^{\varepsilon})}{[(\mu_{PN}^{\varepsilon} - n\sigma_{PN}^{\varepsilon}) - (\mu_{BN}^{\varepsilon} - n\sigma_{BN}^{\varepsilon})] + [(\mu_{BP}^{\varepsilon} - n\sigma_{BP}^{\varepsilon}) - (\mu_{PP}^{\varepsilon} - n\sigma_{PP}^{\varepsilon})]},$$

$$\beta^L = \frac{(\mu_{BN}^{\varepsilon} - n\sigma_{BN}^{\varepsilon}) - (\mu_{NN}^{\varepsilon} - n\sigma_{NN}^{\varepsilon})}{[(\mu_{BN}^{\varepsilon} - n\sigma_{BN}^{\varepsilon}) - (\mu_{NN}^{\varepsilon} - n\sigma_{NN}^{\varepsilon})] + [(\mu_{NP}^{\varepsilon} - n\sigma_{NP}^{\varepsilon}) - (\mu_{BP}^{\varepsilon} - n\sigma_{BP}^{\varepsilon})]},$$

$$\gamma^L = \frac{(\mu_{PN}^{\varepsilon} - n\sigma_{PN}^{\varepsilon}) - (\mu_{NN}^{\varepsilon} - n\sigma_{NN}^{\varepsilon})}{[(\mu_{PN}^{\varepsilon} - n\sigma_{PN}^{\varepsilon}) - (\mu_{NN}^{\varepsilon} - n\sigma_{NN}^{\varepsilon})] + [(\mu_{NP}^{\varepsilon} - n\sigma_{NP}^{\varepsilon}) - (\mu_{PP}^{\varepsilon} - n\sigma_{PP}^{\varepsilon})]};$$

$$\alpha^U = \frac{(\mu_{PN}^{\varepsilon} + n\sigma_{PN}^{\varepsilon}) - (\mu_{BN}^{\varepsilon} + n\sigma_{BN}^{\varepsilon})}{[(\mu_{PN}^{\varepsilon} + n\sigma_{PN}^{\varepsilon}) - (\mu_{BN}^{\varepsilon} + n\sigma_{BN}^{\varepsilon})] + [(\mu_{BP}^{\varepsilon} + n\sigma_{BP}^{\varepsilon}) - (\mu_{PP}^{\varepsilon} + n\sigma_{PP}^{\varepsilon})]},$$

$$\beta^U = \frac{(\mu_{BN}^{\varepsilon} + n\sigma_{BN}^{\varepsilon}) - (\mu_{NN}^{\varepsilon} + n\sigma_{NN}^{\varepsilon})}{[(\mu_{BN}^{\varepsilon} + n\sigma_{BN}^{\varepsilon}) - (\mu_{NN}^{\varepsilon} + n\sigma_{NN}^{\varepsilon})] + [(\mu_{NP}^{\varepsilon} + n\sigma_{NP}^{\varepsilon}) - (\mu_{BP}^{\varepsilon} + n\sigma_{BP}^{\varepsilon})]},$$

$$\gamma^U = \frac{(\mu_{PN}^{\varepsilon} + n\sigma_{PN}^{\varepsilon}) - (\mu_{NN}^{\varepsilon} + n\sigma_{NN}^{\varepsilon})}{[(\mu_{PN}^{\varepsilon} + n\sigma_{PN}^{\varepsilon}) - (\mu_{NN}^{\varepsilon} + n\sigma_{NN}^{\varepsilon})] + [(\mu_{NP}^{\varepsilon} + n\sigma_{NP}^{\varepsilon}) - (\mu_{PP}^{\varepsilon} + n\sigma_{PP}^{\varepsilon})]};$$

Analogously, if $\mu_{\bullet\bullet}^{\varepsilon} + n\sigma_{\bullet\bullet}^{\varepsilon}, \mu_{\bullet\bullet}^{\varepsilon} - n\sigma_{\bullet\bullet}^{\varepsilon}$ satisfy the stricter conditions:

$$\mu_{PP}^{\varepsilon} - n\sigma_{PP}^{\varepsilon} < \mu_{PP}^{\varepsilon} + n\sigma_{PP}^{\varepsilon} \leq \mu_{BP}^{\varepsilon} - n\sigma_{BP}^{\varepsilon}$$
$$< \mu_{BP}^{\varepsilon} + n\sigma_{BP}^{\varepsilon} < \mu_{NP}^{\varepsilon} - n\sigma_{NP}^{\varepsilon} < \mu_{NP}^{\varepsilon} + n\sigma_{NP}^{\varepsilon};$$

$$\mu_{NN}^{\varepsilon} - n\sigma_{NN}^{\varepsilon} < \mu_{NN}^{\varepsilon} + n\sigma_{NN}^{\varepsilon} \leq \mu_{BN}^{\varepsilon} - n\sigma_{BN}^{\varepsilon}$$
$$< \mu_{BN}^{\varepsilon} + n\sigma_{BN}^{\varepsilon} < \mu_{PN}^{\varepsilon} - n\sigma_{PN}^{\varepsilon} < \mu_{PN}^{\varepsilon} + n\sigma_{PN}^{\varepsilon}. \tag{17}$$

The minimum and maximum values for (α, β) by constraining all α and β in the interval $[0, 1]$.

$$\alpha^{min} = \frac{(\mu_{PN}^{\varepsilon} - n\sigma_{PN}^{\varepsilon}) - (\mu_{BN}^{\varepsilon} + n\sigma_{BN}^{\varepsilon})}{[(\mu_{PN}^{\varepsilon} + n\sigma_{PN}^{\varepsilon}) - (\mu_{BN}^{\varepsilon} - n\sigma_{BN}^{\varepsilon})] + [(\mu_{BP}^{\varepsilon} + n\sigma_{BP}^{\varepsilon}) - (\mu_{PP}^{\varepsilon} - n\sigma_{PP}^{\varepsilon})]},$$

$$\alpha^{max} = min(\frac{(\mu_{PN}^{\varepsilon} + n\sigma_{PN}) - (\mu_{BN}^{\varepsilon} - n\sigma_{BN}^{\varepsilon})}{[(\mu_{PN}^{\varepsilon} - n\sigma_{PN}^{\varepsilon}) - (\mu_{BN}^{\varepsilon} + n\sigma_{BN}^{\varepsilon})] + [(\mu_{BP}^{\varepsilon} - n\sigma_{BP}^{\varepsilon}) - (\mu_{PP}^{\varepsilon} + n\sigma_{PP}^{\varepsilon})]}, 1);$$

$$\beta^{min} = \frac{(\mu_{BN}^{\varepsilon} - n\sigma_{BN}^{\varepsilon}) - (\mu_{NN}^{\varepsilon} + n\sigma_{NN}^{\varepsilon})}{[(\mu_{BN}^{\varepsilon} + n\sigma_{BN}^{\varepsilon}) - (\mu_{NN}^{\varepsilon} - n\sigma_{NN}^{\varepsilon})] + [(\mu_{NP}^{\varepsilon} + n\sigma_{NP}^{\varepsilon}) - (\mu_{BP}^{\varepsilon} - n\sigma_{BP}^{\varepsilon})]},$$

$$\beta^{max} = min(\frac{(\mu_{BN}^{\varepsilon} + n\sigma_{BN}^{\varepsilon}) - (\mu_{NN}^{\varepsilon} - n\sigma_{NN}^{\varepsilon})}{[(\mu_{BN}^{\varepsilon} - n\sigma_{BN}^{\varepsilon}) - (\mu_{NN}^{\varepsilon} + n\sigma_{NN}^{\varepsilon})] + [(\mu_{NP}^{\varepsilon} - n\sigma_{NP}^{\varepsilon}) - (\mu_{BP}^{\varepsilon} + n\sigma_{BP}^{\varepsilon})]}, 1).$$

where, $n = 1, 2, 3$ corresponds to one of the three different degrees of confidence. Obviously, $(\alpha_L, \alpha_U) \subseteq [\alpha^{min}, \alpha^{max}]$, $(\beta_L, \beta_U) \subseteq [\beta^{min}, \beta^{max}]$. However, in many cases, (17) may not be valid. In this scenario, one should find the proper n^* satisfies that $\mu_{PP}^{\varepsilon} + n^*\sigma_{PP}^{\varepsilon} \leq \mu_{BP}^{\varepsilon} - n^*\sigma_{BP}^{\varepsilon}$, $\mu_{BP}^{\varepsilon} + n^*\sigma_{BP}^{\varepsilon} < \mu_{NP}^{\varepsilon} - n^*\sigma_{NP}^{\varepsilon}$, $\mu_{NN}^{\varepsilon} + n^*\sigma_{NN}^{\varepsilon} \leq \mu_{BN}^{\varepsilon} - n^*\sigma_{BN}^{\varepsilon}$, $\mu_{BN}^{\varepsilon} + n^*\sigma_{BN}^{\varepsilon} < \mu_{PN}^{\varepsilon} - n^*\sigma_{PN}^{\varepsilon}$.

Note that, the "68-95-99.7" rule is one special case of (6) in Sect. 3.1.1. For example, let the significance level be l=0.1, 0.05 and 0.01, we have n= 1.64, 1.96 and 2.58, respectively. Observed by the fact that the width of the interval $[\mu_{\bullet\bullet}^{\varepsilon} - n\sigma_{\bullet\bullet}^{\varepsilon}, \mu_{\bullet\bullet}^{\varepsilon} + n\sigma_{\bullet\bullet}^{\varepsilon}]$ is depended on n, one should choose a proper n to tradeoff the margin of error and the requirement of (17) in real decision problems.

4   An Illustration

In this section, we illustrate the SDTRS model by an example of decision in Public-Private Partnerships (PPP) project investment. A PPP project is funded and operated through a partnership of government and one or more private sectors according to their contract. Through a PPP project, the government can reduces financial expenditure and sufficiently allocates the limited recourse, and the private sectors can benefit from the PPP project by using their technology, fund and professional knowledge. However, because of the implementation of PPP project, there exist many risk and uncertain factors, e.g., design, construction, and maintenance risk; demand/revenue risk, political risk, etc.

During the risk assessment of PPP project investment, we have two states $\Omega = \{C, \neg C\}$ indicating that the project is a good project and bad project, respectively. With respect to the three-way decision, the set of actions is given by $\mathcal{A} = \{a_P, a_B, a_N\}$, where a_P, a_B, and a_N represent investment, need further analysis and do not investment, respectively. There are 6 parameters in the model. $\lambda_{PP}^{\varepsilon}$, $\lambda_{BP}^{\varepsilon}$ and $\lambda_{NP}^{\varepsilon}$ denote the costs incurred for taking actions of investment, need further analysis and do not investment when a PPP project is good; $\lambda_{PN}^{\varepsilon}$, $\lambda_{BN}^{\varepsilon}$ and $\lambda_{NN}^{\varepsilon}$ denote the costs incurred for taking actions of investment,

Table 1. The losses of 6 types of PPP projects

X	$\lambda^\varepsilon_{\bullet\bullet} \sim U(a,b)$											
	a_{PP}	b_{PP}	a_{BP}	b_{BP}	a_{NP}	b_{NP}	a_{PN}	b_{PN}	a_{BN}	b_{BN}	a_{NN}	b_{NN}
x_1	0.5	1.5	1	3	2	6	2.5	7.5	1.5	4.5	1	3
x_2	2	4	4	6	7	9	3	5	0.25	0.75	0	0
x_3	0.5	1.5	1	2	2	4	4	6	0.5	1.5	0	0
x_4	3	5	6	8	10	12	5	9	3	6	1	3
x_5	1	3	2	4	3	5	2	4	0.5	1.5	0	1
x_6	0	2	1	3	3	4	4	6	2	4	1	3

X	$\lambda^\varepsilon_{\bullet\bullet} \sim N(\mu,\sigma^2)$											
	μ^ε_{PP}	σ^ε_{PP}	μ^ε_{BP}	σ^ε_{BP}	μ^ε_{NP}	σ^ε_{NP}	μ^ε_{PN}	σ^ε_{PN}	μ^ε_{BN}	σ^ε_{BN}	μ^ε_{NN}	σ^ε_{NN}
x_1	1	0.1	2	0.2	4	0.25	5	0.3	3	0.15	2	0.05
x_2	3	0.2	5	0.4	8	0.6	4	0.3	0.5	0.05	0	0
x_3	1	0.05	1.5	0.1	3	0.2	5	0.5	1	0.05	0	0
x_4	4	0.2	7	0.3	11	0.5	7	0.3	4.5	0.2	2	0.1
x_5	2	0.1	3	0.1	4	0.1	3	0.2	1	0.1	0.5	0
x_6	1	0.05	2	0.1	3.5	0.15	5	0.2	3	0.1	2	0.05

need further analysis and do not investment when a PPP project is bad. Here, we consider two scenarios: (1) $\lambda^\varepsilon_{\bullet\bullet}$ obeys uniform distribution; (2) $\lambda^\varepsilon_{\bullet\bullet}$ obeys normal distribution. Table 1 shows the losses for six types of PPP projects.

Based on Table 1 and our discussions in Sect. 3.2, the three thresholds for each PPP project can be obtained. In the following, we consider $\lambda^\varepsilon_{\bullet\bullet} \sim U(a,b)$ and $\lambda^\varepsilon_{\bullet\bullet} \sim N(\mu,\sigma^2)$ ($n = 1,2,3$). Figure 1 outlines the calculating results of thresholds α, β and γ for each PPP investment project when $\lambda^\varepsilon_{\bullet\bullet} \sim U(a,b)$. Table 2 lists the calculating results of thresholds α, β and γ for each PPP investment project when $\lambda^\varepsilon_{\bullet\bullet} \sim N(\mu,\sigma^2)$ ($n = 1,2,3$).

In Fig. 1 and Table 2, it can be seen that different pairs of thresholds are obtained for different types of PPP projects. For each $x_i \in X$ ($i = 1,2,\cdots,6$), $\alpha^L, \alpha^U \in [\alpha^{min}, \alpha^{max}]$ and $\beta^L, \beta^U \in [\beta^{min}, \beta^{max}]$. Suppose that $Pr(X|x_i) = 0.5$ for all $x_i \in X$. By three-way decision rules (P1), (B1) and (N1), we compare the thresholds in Fig. 1 and Table 2 and the conditional probability $Pr(X|x_i)$ for each x_i, then list the decision regions in Table 3 by considering different α, β and γ.

In Table 3, given a pair of (α, β), the government can approve the project x_i immediately with $x_i \in POS(X)$, and they should give up to implement the PPP project x_i immediately with $x_i \in NEG(X)$. In addition, the PPP project x_i needs to be further investigated with $x_i \in BND(X)$, which the government should delay to make a decision. In conclusion, given a PPP project x_i, we simply compare the relation between the conditional probability $Pr(X|x_i)$ and the thresholds (α, β) to make a decision.

Table 2. The values of three parameters for each PPP investment project when $\lambda_{\bullet\bullet}^{\xi} \sim N(\mu, \sigma^2)$ $(n = 1, 2, 3)$

X	α^{min}	β^{min}	α^{max}	β^{max}	α^L	β^L	γ^L	α^U	β^U	γ^U
	Normal Distribution $(n=1)$									
x_1	0.4133	0.2051	1.0000	0.5106	0.6727	0.3158	0.4911	0.6615	0.3492	0.5078
x_2	0.4884	0.0874	0.8462	0.2245	0.6436	0.1385	0.4458	0.6303	0.1467	0.4433
x_3	0.6635	0.3115	1.0000	0.4884	0.8875	0.4043	0.7087	0.8900	0.3962	0.7190
x_4	0.3077	0.2716	0.6667	0.5185	0.4528	0.3871	0.4174	0.4561	0.3824	0.4160
x_5	0.4857	0.2105	0.9200	0.5000	0.6552	0.2857	0.5349	0.6774	0.3750	0.5745
x_6	0.4928	0.2787	0.9020	0.5476	0.6667	0.3958	0.5429	0.6667	0.4038	0.5478
	Normal Distribution $(n=2)$									
x_1	0.2444	0.1395	1.0000	0.8235	0.6800	0.2963	0.4808	0.6571	0.3636	0.5147
x_2	0.3784	0.0714	1.0000	0.4286	0.6522	0.1333	0.4474	0.6250	0.1500	0.4423
x_3	0.4915	0.2813	1.0000	0.6111	0.8857	0.4091	0.7018	0.8909	0.3929	0.7229
x_4	0.2000	0.2184	1.0000	0.7209	0.4510	0.3898	0.4182	0.4576	0.3803	0.4154
x_5	0.3500	0.1429	1.0000	0.7778	0.6429	0.2308	0.5122	0.6875	0.4118	0.5918
x_6	0.3590	0.2121	1.0000	0.7647	0.6667	0.3913	0.5400	0.6667	0.4074	0.5500
	Normal Distribution $(n=3)$									
x_1	0.1238	0.0808	1.0000	1.0000	0.6889	0.2745	0.4688	0.6533	0.3768	0.5208
x_2	0.2934	0.0526	1.0000	1.0000	0.6627	0.1273	0.4493	0.6204	0.1529	0.4414
x_3	0.3561	0.2394	1.0000	0.7931	0.8833	0.4146	0.6931	0.8917	0.3898	0.7263
x_4	0.1176	0.1633	1.0000	1.0000	0.4490	0.3929	0.4190	0.4590	0.3784	0.4148
x_5	0.2444	0.0833	1.0000	1.0000	0.6296	0.1667	0.4872	0.6970	0.4444	0.6078
x_6	0.2529	0.1486	1.0000	1.0000	0.6667	0.3864	0.5368	0.6667	0.4107	0.5520

Table 3. The decision region for each PPP projects when $Pr(X|x_i) = 0.5$

Distribution	(α, β)	$POS(X)$	$BND(X)$	$NEG(X)$
Uniform	$(\overline{\alpha}, \overline{\beta})$	$\{x_4\}$	$\{x_1, x_2, x_3, x_5, x_6\}$	$\varnothing$
Normal $(n=1)$	$(\alpha^{min}, \beta^{min})$	$\{x_1, x_2, x_4, x_5, x_6\}$	$\{x_3\}$	$\varnothing$
	$(\alpha^{max}, \beta^{max})$	$\varnothing$	$\{x_2, x_3\}$	$\{x_1, x_4, x_5, x_6\}$
	(α^L, β^L)	$\{x_4\}$	$\{x_1, x_2, x_3, x_5, x_6\}$	$\varnothing$
	(α^U, β^U)	$\{x_4\}$	$\{x_1, x_2, x_3, x_5, x_6\}$	$\varnothing$
Normal $(n=2)$	$(\alpha^{min}, \beta^{min})$	X	$\varnothing$	$\varnothing$
	$(\alpha^{max}, \beta^{max})$	$\varnothing$	$\{x_2\}$	$\{x_1, x_3, x_4, x_5, x_6\}$
	(α^L, β^L)	$\{x_4\}$	$\{x_1, x_2, x_3, x_5, x_6\}$	$\varnothing$
	(α^U, β^U)	$\{x_4\}$	$\{x_1, x_2, x_3, x_5, x_6\}$	$\varnothing$
Normal $(n=3)$	$(\alpha^{min}, \beta^{min})$	X	$\varnothing$	$\varnothing$
	$(\alpha^{max}, \beta^{max})$	$\varnothing$	$\varnothing$	X
	(α^L, β^L)	$\{x_4\}$	$\{x_1, x_2, x_3, x_5, x_6\}$	$\varnothing$
	(α^U, β^U)	$\{x_4\}$	$\{x_1, x_2, x_3, x_5, x_6\}$	$\varnothing$

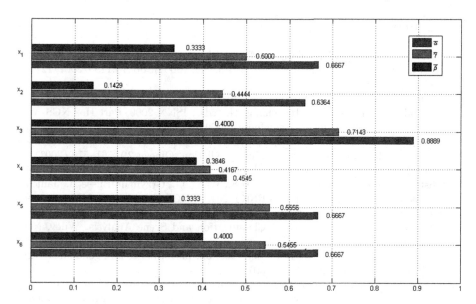

Fig. 1. The values of three parameters for each PPP investment project when $\lambda_{\bullet\bullet}^{\varepsilon} \sim U(a, b)$

5 Conclusions

As one of important mathematical methodologies to express the uncertainty, a stochastic approach is one in which values are obtained from a corresponding sequence of jointly distributed random variables. In this paper, we introduced the "confidence interval" method to DTRS, and proposed SDTRS in which the stochastic loss functions are estimated by "confidence interval". Furthermore, two extension of SDTRS models were proposed by considering the loss functions obey uniform distribution and normal distribution, respectively. The probabilistic distribution elicitation of stochastic number provides a solution to obtain loss function values for SDTRS model. An example of PPP project investment was utilized to illustrate our approach. Our future research work will focus on developing extended DTRS models in other uncertain decision environments.

Acknowledgements. This paper is an extended version of the paper published in the proceedings of RSKT2012. This work is partially supported by the National Science Foundation of China (Nos. 71201133, 61175047, 71090402, 71201076), the Youth Social Science Foundation of the Chinese Education Commission (No. 11YJC630127), the Research Fund for the Doctoral Program of Higher Education of China (No. 20120184120028), the China Postdoctoral Science Foundation (Nos. 2012M520310, 2013T60132) and the Fundamental Research Funds for the Central Universities of China (No. SWJTU12CX117).

References

1. Abd El-Monsef, M.M.E., Kilany, N.M.: Decision analysis via granulation based on general binary relation. Int. J. Math. Math. Sci. (2007). doi:10.1155/2007/12714. Article ID 12714
2. Abbas, A.R., Juan, L.: Supporting e-learning system with modified Bayesian rough set model. In: Yu, W., He, H., Zhang, N. (eds.) ISNN 2009, Part II. LNCS, vol. 5552, pp. 192–200. Springer, Heidelberg (2009)
3. Beynon, M.J., Peel, M.J.: Variable precision rough set theory and data discretisation: an application to corporate failure prediction. Omega **29**, 561–576 (2001)
4. Deng, X., Yao, Y.: An information-theoretic interpretation of thresholds in probabilistic rough sets. In: Li, T., Nguyen, H.S., Wang, G., Grzymala-Busse, J., Janicki, R., Hassanien, A.E., Yu, H. (eds.) RSKT 2012. LNCS (LNAI), vol. 7414, pp. 369–378. Springer, Heidelberg (2012)
5. Duda, R.O., Hart, P.E.: Pattern Classification and Scene Analysis. Wiley, New York (1973)
6. Forster, M.R.: Key concepts in model selection: performance and generalizability. J. Math. Psychol. **44**, 205–231 (2000)
7. Goudey, R.: Do statistical inferences allowing three alternative decision give better feedback for environmentally precautionary decision-making. J. Environ. Manag. **85**, 338–344 (2007)
8. Herbert, J.P., Yao, J.T.: Game-theoretic rough sets. Fundam. Inform. **108**, 267–286 (2011)
9. Jia, X.Y., Liao, W.H., Tang, Z.M., Shang, L.: Minimum cost attribute reduction in decision-theoretic rough set models. Inf. Sci. **219**, 151–167 (2012)
10. Li, H.X., Zhou, X.Z.: Risk decision making based on decision-theoretic rough set: a three-way view decision model. Int. J. Comput. Intell. Syst. **4**, 1–11 (2011)
11. Li, H.X., Zhou, X.Z., Zhao, J.B., Liu, D.: Non-monotonic attribute reduction in decision-theoretic rough sets. Fundam. Inform. **126**, 415–432 (2013)
12. Liang, D.C., Liu, D., Pedrycz, W., Hu, P.: Triangular fuzzy decision-theoretic rough sets. Int. J. Approx. Reason. **54**, 1087–1106 (2013)
13. Lingras, P., Chen, M., Miao, D.Q.: Rough cluster quality index based on decision theory. IEEE Trans. Knowl. Data Eng. **21**, 1014–1026 (2009)
14. Lingras, P., Chen, M., Miao, D.Q.: Semi-supervised rough cost/benefit decisions. Fundam. Inform. **94**, 233–244 (2009)
15. Liu, D., Li, H.X., Zhou, X.Z.: Two decades' research on decision-theoretic rough sets. In: Proceedings of 9th IEEE International Conference on Cognitive Informatics, pp. 968–973 (2010)
16. Liu, D., Yao, Y.Y., Li, T.R.: Three-way investment decisions with decision-theoretic rough sets. Int. J. Comput. Intell. Syst. **4**, 66–74 (2011)
17. Liu, D., Li, T.R., Ruan, D.: Probabilistic model criteria with decision-theoretic rough sets. Inf. Sci. **181**, 3709–3722 (2011)
18. Liu, D., Li, T., Liang, D.: A new discriminant analysis approach under decision-theoretic rough sets. In: Yao, J.T., Ramanna, S., Wang, G., Suraj, Z. (eds.) RSKT 2011. LNCS (LNAI), vol. 6954, pp. 476–485. Springer, Heidelberg (2011)
19. Liu, D., Li, T.R., Li, H.X.: A multiple-category classification approach with decision-theoretic rough sets. Fundam. Inform. **115**, 173–188 (2012)
20. Liu, D., Li, T., Liang, D.: Decision-theoretic rough sets with probabilistic distribution. In: Li, T., Nguyen, H.S., Wang, G., Grzymala-Busse, J., Janicki, R., Hassanien, A.E., Yu, H. (eds.) RSKT 2012. LNCS (LNAI), vol. 7414, pp. 389–398. Springer, Heidelberg (2012)

21. Liu, D., Li, T.R., Liang, D.C.: Three-way government decision analysis with decision-theoretic rough sets. Int. J. Uncertain. Fuzziness Knowl. Based Syst. **20**, 119–132 (2012)
22. Liu, D., Li, T.R., Li, H.X.: Interval-valued decision-theoretic rough sets. Comput. Sci. **39**(7), 178–181 (2012). (in chinese)
23. Liu, D., Li, T.R., Liang, D.C.: Fuzzy decision-theoretic rough sets. Comput. Sci. **39**(12), 25–29 (2012). (in chinese)
24. Liu, D., Li, T.R., Liang, D.C.: Incorporating logistic regression to decision-theoretic rough sets for classifications. Int. J. Approx. Reason. **55**, 197–210 (2014)
25. Liu, D., Li, T.R., Liang, D.C.: Fuzzy interval decision-theoretic rough sets. In: Proceedings of 2013 IFSA World Congress NAFIPS Annual Meeting, pp. 1315–1320 (2013)
26. Ma, W.M.: On relationship between probabilistic rough set and Bayesian risk decision over two universes. Int. J. Gen. Syst **41**, 225–245 (2012)
27. Macmillan, F.: Risk, uncertainty and investment decision-making in the upstream oil and gas industry. Ph.D thesis, University of Aberdeen, UK (2000)
28. Min, F., He, H.P., Qian, Y.H., Zhu, W.: Test-cost-sensitive attribute reduction. Inf. Sci. **181**, 4928–4942 (2011)
29. Min, F., Zhu, W.: Attribute reduction of data with error ranges and test costs. Inf. Sci. **211**, 48–67 (2012)
30. Mishra, A., Mishra, M., Shiv, B.: In praise of vagueness: malleability of vague information as a performance-booster. Psychol. Sci. **22**, 733–738 (2012)
31. Pauker, S.G., Kassirer, J.P.: The threshold approach to clinical decision making. N. Engl. J. Med. **302**, 1109–1117 (1980)
32. Pawlak, Z.: Rough sets. Int. J. Comput. Inf. Sci. **11**, 341–356 (1982)
33. Pawlak, Z., Wong, S.K.M., Ziarko, W.: Rough sets: probabilistic versus deterministic approach. Int. J. Man Mach. Stud. **29**, 81–95 (1988)
34. Pawlak, Z.: Rough Wets, Theoretical Aspects of Reasoning About Data. Kluwer Academic Publishers Press, Dordrecht (1991)
35. Skowron, A., Stepaniuk, J.: Tolerance approximation spaces. Fundam. Inform. **27**, 245–253 (1996)
36. Ślęzak, D.: Rough sets and Bayes factor. In: Peters, J.F., Skowron, A. (eds.) Transactions on Rough Sets III. LNCS, vol. 3400, pp. 202–229. Springer, Heidelberg (2005)
37. Ślęzak, D., Ziarko, W.: The investigation of the Bayesian rough set model. Int. J. Approx. Reason. **40**, 81–91 (2005)
38. Ślęzak, D., Wróblewski, J., Eastwood, V., Synak, P.: Brighthouse: an analytic data warehouse for ad-hoc queries. Proc. VLDB Endow. **1**, 1337–1345 (2008)
39. Woodward, P.W., Naylor, J.C.: An application of Bayesian methods in SPC. The Statistician **42**, 461–469 (1993)
40. Xie, G., Yue, W., Wang, S.Y., Lai, K.K.: Dynamic risk management in petroleum project investment based on a variable precision rough set model. Technol. Forecast. Soc. Chang. **77**, 891–901 (2010)
41. Yang, X.P., Yao, J.T.: Modelling multi-agent three-way decisions with decision-theoretic rough sets. Fundam. Inform. **115**, 157–171 (2012)
42. Yao, J.T., Herbert, J.P.: Web-based support systems with rough set analysis. In: Kryszkiewicz, M., Peters, J.F., Rybiński, H., Skowron, A. (eds.) RSEISP 2007. LNCS (LNAI), vol. 4585, pp. 360–370. Springer, Heidelberg (2007)
43. Yao, Y.Y., Wong, S.K.M., Lingras, P.: A decision-theoretic rough set model. In: Proceedings of the 5th International Symposium on Methodologies for Intelligent Systems, pp. 17–25 (1990)

44. Yao, Y.Y., Wong, S.K.M.: A decision theoretic framework for approximating concepts. Int. J. Man Mach. Stud. **37**, 793–809 (1992)
45. Yao, Y.Y.: Probabilistic approaches to rough sets. Expert Syst. **20**, 287–297 (2003)
46. Yao, Y.: Decision-theoretic rough set models. In: Yao, J.T., Lingras, P., Wu, W.-Z., Szczuka, M.S., Cercone, N.J., Ślęzak, D. (eds.) RSKT 2007. LNCS (LNAI), vol. 4481, pp. 1–12. Springer, Heidelberg (2007)
47. Yao, Y.Y.: Probabilistic rough set approximations. Int. J. Approx. Reason. **49**, 255–271 (2008)
48. Yao, Y.Y., Zhao, Y.: Attribute reduction in decision-theoretic rough set models. Inf. Sci. **178**, 3356–3373 (2008)
49. Yao, Y.: Three-way decision: an interpretation of rules in rough set theory. In: Wen, P., Li, Y., Polkowski, L., Yao, Y., Tsumoto, S., Wang, G. (eds.) RSKT 2009. LNCS, vol. 5589, pp. 642–649. Springer, Heidelberg (2009)
50. Yao, Y.Y.: Three-way decisions with probabilistic rough sets. Inf. Sci. **180**, 341–353 (2010)
51. Yao, Y., Zhou, B.: Naive Bayesian rough sets. In: Yu, J., Greco, S., Lingras, P., Wang, G., Skowron, A. (eds.) RSKT 2010. LNCS (LNAI), vol. 6401, pp. 719–726. Springer, Heidelberg (2010)
52. Yao, Y.Y.: The superiority of three-way decision in probabilistic rough set models. Inf. Sci. **181**, 1080–1096 (2011)
53. Yao, Y.Y.: Two semantic issues in a probabilistic rough set model. Fundam. Inform. **108**, 249–265 (2011)
54. Yao, Y., Zhou, B.: Naive Bayesian Rough Sets. In: Yu, J., Greco, S., Lingras, P., Wang, G., Skowron, A. (eds.) RSKT 2010. LNCS, vol. 6401, pp. 719–726. Springer, Heidelberg (2010)
55. Yao, Y.: An outline of a theory of three-way decisions. In: Yao, J.T., Yang, Y., Słowiński, R., Greco, S., Li, H., Mitra, S., Polkowski, L. (eds.) RSCTC 2012. LNCS (LNAI), vol. 7413, pp. 1–17. Springer, Heidelberg (2012)
56. Yusgiantoro, P., Hsiao, F.S.T.: Production-sharing contracts and decision-making in oil production. Energy Econ. **10**, 245–256 (1993)
57. Zhao, W.Q., Zhu, Y.L.: An email classification scheme based on decision-theoretic rough set theory and analysis of email security. In: Proceedings of 2005 IEEE Region 10 TENCON, pp. 1–6 (2005)
58. Zhao, Y., Wong, S.K.M., Yao, Y.: A note on attribute reduction in the decision-theoretic rough set model. In: Peters, J.F., Skowron, A., Chan, C.-C., Grzymala-Busse, J.W., Ziarko, W.P. (eds.) Transactions on Rough Sets XIII. LNCS, vol. 6499, pp. 260–275. Springer, Heidelberg (2011)
59. Zhou, B., Yao, Y.Y., Luo, J.G.: A three-way decision approach to email spam filtering. In: Proceedings of 23th Canadian AI, pp. 28–39 (2010)
60. Zhou, B.: A new formulation of multi-category decision-theoretic rough sets. In: Yao, J.T., Ramanna, S., Wang, G., Suraj, Z. (eds.) RSKT 2011. LNCS (LNAI), vol. 6954, pp. 514–522. Springer, Heidelberg (2011)
61. Ziarko, W.: Variable precision rough set model. J. Comput. Syst. Sci. **46**, 39–59 (1993)
62. Ziarko, W.: Probabilistic approach to rough set. Int. J. Approx. Reason. **49**, 272–284 (2008)

Author Index

Printed in the United States
By Bookmasters